ADVANCES IN
CELL BIOLOGY

Volume 1 • 1987

ADVANCES IN CELL BIOLOGY

A Research Annual

Editor: KENNETH R. MILLER
Division of Biology and Medicine
Brown University

VOLUME 1 • 1987

JAI PRESS INC.

Greenwich, Connecticut *London, England*

55 Old Post Road, No. 2
Greenwich, Connecticut 06836

JAI PRESS LTD.
3 Henrietta Street
London WC2E 8LU
England

ISBN: 0-89232-792-8

Manufactured in the United States of America

CONTENTS

LIST OF CONTRIBUTORS

Robert A. Bloodgood	Department of Anatomy and Cell Biology University of Virginia
David R. Burgess	Department of Anatomy and Cell Biology University of Miami
Knute A. Fisher	Department of Biochemistry and Biophysics and Cardiovascular Research Institute University of California San Francisco
Kenneth R. Miller	Division of Biology and Medicine Brown University
Phyllis M. Novikoff	Department of Pathology Yeshiva University
Pedro Pinto da Silva	Laboratory of Mathematical Biology National Cancer Institute Frederick, MD

INTRODUCTION

In the formal sense, cell biology is one of the youngest of biology's many subfields. International meetings on cell biology, formal organizations, and journals of cell biology are comparatively recent entries into the bewildering thicket of scientific disciplines. Yet cell biology is one of the oldest and most fundamental routes of inquiry into the nature of living systems. A justification for emphasizing the importance of cell biology is easy enough to develop. The cell is the basic unit of life, and detailed studies of cells, their structure, behavior, and their components are clearly required to develop a detailed appreciation of living organisms.

The techniques of research in cell biology are as varied as the imaginations of the investigators who still find the cell to be an intriguing target. From the original tool of cell biology, the light microscope, researchers have developed a toolbench that includes biological, biophysical, chemical, and molecular approaches to the understanding of cell structure and cell function. One of the reasons for the great vitality of cell biology at this period in the history of biology is precisely this diversity of approach and technique, and techniques developed for use in cell biology are now beginning to have wider applications in other fields of biology, medicine, and biotechnology.

In the series that is inaugurated with this volume we hope to capture some of the excitement and vitality that characterizes the state of the field today. We have chosen an approach in which individual researchers at the forefront of cell biology have been invited to highlight some of the most important research and technical advances in their research. While these articles will integrate these advances into the large picture of cell biology research, the authors have not been burdened with the task of reviewing all advances in their fields or assembling an exhaustive citation list from which the reader can develop a bibliography of all work published on a particular topic. Instead, we have tried to choose a format in which individual researchers will do what they do best: to explain the areas in which they work and to highlight the research developments and advances which are shaping that field and influencing the direction of scientific thought and experimentation. In doing this, I hope that we will present to the scientific public a series that will improve communication and dialogue, and will bring to cell biology a focused annual series containing the scientific highlights of the past year.

Kenneth R. Miller
Series Editor

THE ANALYSIS OF FRACTURED MEMBRANE "HALVES"

Knute A. Fisher

I. INTRODUCTION

The physical and chemical analysis of split membrane leaflets is based on several common principles and observations. First, it has been established that the process of freeze fracture cleaves the lipid bilayer of the membrane, separating the inner and outer lipid leaflets (Branton, 1966; Pinto da Silva and Branton, 1970). Second, the split leaflets or "halves," like intact membranes, are known to be composed of protein, lipid, and carbohydrate. Third, transmission electron microscopy (TEM) consistently reveals intramembranous particles (IMPs) in all biological membranes.[1] And finally, it is assumed that IMPs represent membrane-spanning polypeptides and/or lipid-polypeptide aggregates and that the surrounding smooth regions represent lipid monolayers.

Physical analyses of fractured membrane "halves" are usually based on electron microscopy. Freeze-fractured samples are shadowed with electron-

Advances in Cell Biology, Vol. 1, pages 1–29.

ISBN: 0-89232-792-8

scattering metals such as platinum-carbon, tungsten, or tantalum, and the surface replica is examined by TEM. Chemical information about split membranes is usually derived by correlation or by cytochemistry. Correlative studies include those in which treated and control cell membranes are examined for changes in IMP number or distribution. If the transmembrane or in-plane distribution of IMPs is found to be altered by the treatment, the morphological changes are often explained in molecular terms. More direct cytochemical approaches have also been developed, including methods to label lipids, carbohydrates, and polypeptides. For example, polyene antibiotics such as Filipin have been used to evaluate the cholesterol content of freeze-fractured membranes (Verkleij et al., 1973; Elias et al. 1979). Filipin combines with unesterified 3-β-hydroxysterols and forms large protuberances and pits in the fractured membranes. Labels for anionic lipids have also been developed (Bearer and Friend, 1980), and membrane carbohydrates are routinely investigated by lectin labels conjugated to microscopic markers such as ferritin or gold.

IMP labeling methods were first developed in the early 1970s to validate membrane splitting (Pinto da Silva and Branton, 1970), and they have been used to probe the surface distribution of lectin (Pinto da Silva and Nicolson, 1974) and antibody binding sites (Pinto da Silva et al., 1971). The distribution of the marker on the etched surface can be correlated with surface morphology and with the distribution of IMPs on contiguous fracture faces. In erythrocyte ghost membrane studies, pH-induced aggregation of IMPs aided the localization of ferritin conjugates relative to the particle clusters (Pinto da Silva, 1972).

Over the past several years the resolution of the labeling approaches has increased significantly. Now, in principle, individual IMPs can be labeled. In a recent method called "fracture-label" (Pinto da Silva et al., 1981), chemically and physically stabilized membranes are fractured, thawed, labeled with colloidal gold or ferritin conjugates, and processed for thin-section TEM. In the "sectioned labelled-replica technique" (Rash et al., 1982), replicated and labeled tissues are processed for thin-section TEM. In a more recent technique, dubbed "label-fracture" (Pinto da Silva and Kan, 1984), colloidal gold protein conjugates are used to label the outer surfaces (ES) of the membrane before fracturing. After freeze fracturing and shadowing, the position of the underlying gold relative to the IMPs can be evaluated. These methods are covered in some detail by Pinto da Silva (later chapter in this volume) and will not be discussed here further.

This brief review is focused on an approach to the analysis of fractured membranes that avoids chemical fixation and thus allows the examination of both native and probe molecules. This approach involves the preparation of highly oriented membrane monolayers by adsorption to planar surfaces. The planar membrane monolayers are freeze fractured, and the split membrane fractions are analyzed either by electron microscopic autoradiography or by direct physical and chemical methods.

Several simple concepts and basic assumptions provide the foundation for planar membrane splitting. Electrostatic attachment of cells and membranes to a surface is based on the principle of attraction between the negatively charged cell surface and a positively charged support, e.g., polylysine-treated glass (Nevo et al., 1955; Fisher, 1975; Mazia et al., 1975) or Alcian Blue-treated mica (Sommer, 1977; Nermut, 1981). Cells have also been attached by aldehydes (Büechi and Bächi, 1979) or lectins (Alpin and Hughes, 1981), but such procedures will not be reviewed here. The formation of planar cell and membrane monolayers is based on the testable assumption that washing attached cells or shearing or sonicating attached membranes will remove unattached material. Central to planar membrane splitting is the assumption that the process of freeze fracture physically separates the two leaflets of the lipid bilayer (Branton, 1966). And, finally, the possibility for "half" membrane enrichment and isolation was derived from the observation that planar membranes lying parallel to the plane of fracture will fracture in preference to those oriented randomly (Fisher, 1975). This review will highlight recent advances in planar monolayer autoradiography and biochemistry and will provide a perspective for the contributions of these studies to the field of cell and membrane biology.

II. TECHNICAL ADVANCES

A. Polylysine-Treated Planar Surfaces

Because most cells and membranes are negatively charged at physiological pH, they will bind to surfaces carrying the opposite charge. Positively charged planar surfaces can be easily prepared by treating clean, negatively charged surfaces with cations. Mazia and co-workers (1975) prepared polylysine-treated, carbon-coated grids for electron microscopy studies. Clean glass, too, provides an excellent surface for adsorption of polylysine (Nevo et al., 1955; Fisher, 1975). Alcian Blue, a tetravalent cationic dye, has also been used to produce a positive charge on glass (Sommer, 1977) or mica (Nermut, 1981) surfaces for both microscopic and chemical studies. For the chemical studies undertaken in my laboratory, however, polylysine-treated glass has been used exclusively because the color and copper ions present in Alcian Blue interfere with some spectrophotometric and biochemical assays.

Planar preparations of cells and membranes are well suited to both spectroscopic and microscopic analyses (Fisher, 1981). In addition, they are useful for biochemical studies of molecular and transmembrane topographies. The orientation and amount of membrane can be directly quantified by microscopy, and attached membranes can be chemically or enzymatically modified with precise control (for reviews see Fisher, 1978, 1980; Nermut, 1982, 1984).

B. Planar Cell Monolayers

A wide variety of cells have been attached to planar surfaces and examined by microscopy. Examples include *Halobacterium halobium* purple membranes (Fisher and Stoeckenius, 1977; Nermut and Williams, 1977), *Saccharomyces cerevisiae* (Maurer and Mühlethaler, 1981), *Dictyostelium discoideum* (Clarke et al., 1975), EL-4 mouse lymphoma cells (Bächi and Büechi, 1981), KB cells (Nermut, 1982), Vero cells (Nermut, 1982), macrophages (Aggeler and Werb, 1982), and red blood cells from humans (Fisher, 1975; Edwards et al. 1979) and sheep (Lang and Nermut, 1980).

Planar cell monolayers (Figure 1a) have been examined by a variety of light and electron microscopic techniques including scanning electron microscopy, thin section TEM, surface shadowing, negative staining, and freeze fracture (Figure 1b). Cell monolayers are especially useful for the enrichment and isolation of the plasma membrane and, in membrane splitting studies, for isolating the outer leaflet of the membrane bilayer. Single membrane monolayers, on the other hand, are especially useful for the enrichment and isolation of inner leaflets.

C. Planar Membrane Monolayers

Planar membrane monolayers are flattened single membranes attached to a planar surface. They are thus composed of both leaflets of the lipid bilayer. Membrane monolayers can be formed using isolated membrane sheets, such as purple membrane, or intact cells, such as human erythrocytes (Figure 2a), given that they have been washed well to remove soluble proteins, which compete for surface binding sites. The formation of bona fide single membrane monolayers (SMM) requires methods that are sample dependent; each new preparation should be verified by light and electron microscopy. For example, single sheets of *H. halobium* purple membrane can be attached, as a function of pH, to polylysine-treated glass (PLG) by either their extracellular (Figure 2b) or cytoplasmic (Figure 2c) surfaces. To remove aggregated, overlapped, or folded membranes, the preparations must be briefly sonicated (Fisher, 1981). In contrast, SMM of attached intact cells, ghosts, or vesicles are usually formed by hydraulic shearing. This process has been used to expose the cytoplasmic surface of the membrane for microscopic examination of cytoskeletal structures (Clarke et al., 1975). Nermut (1982) dubbed his version of the method "lysis-squirting," and I have referred to it as "hydraulic shearing" (Fisher, 1982a).

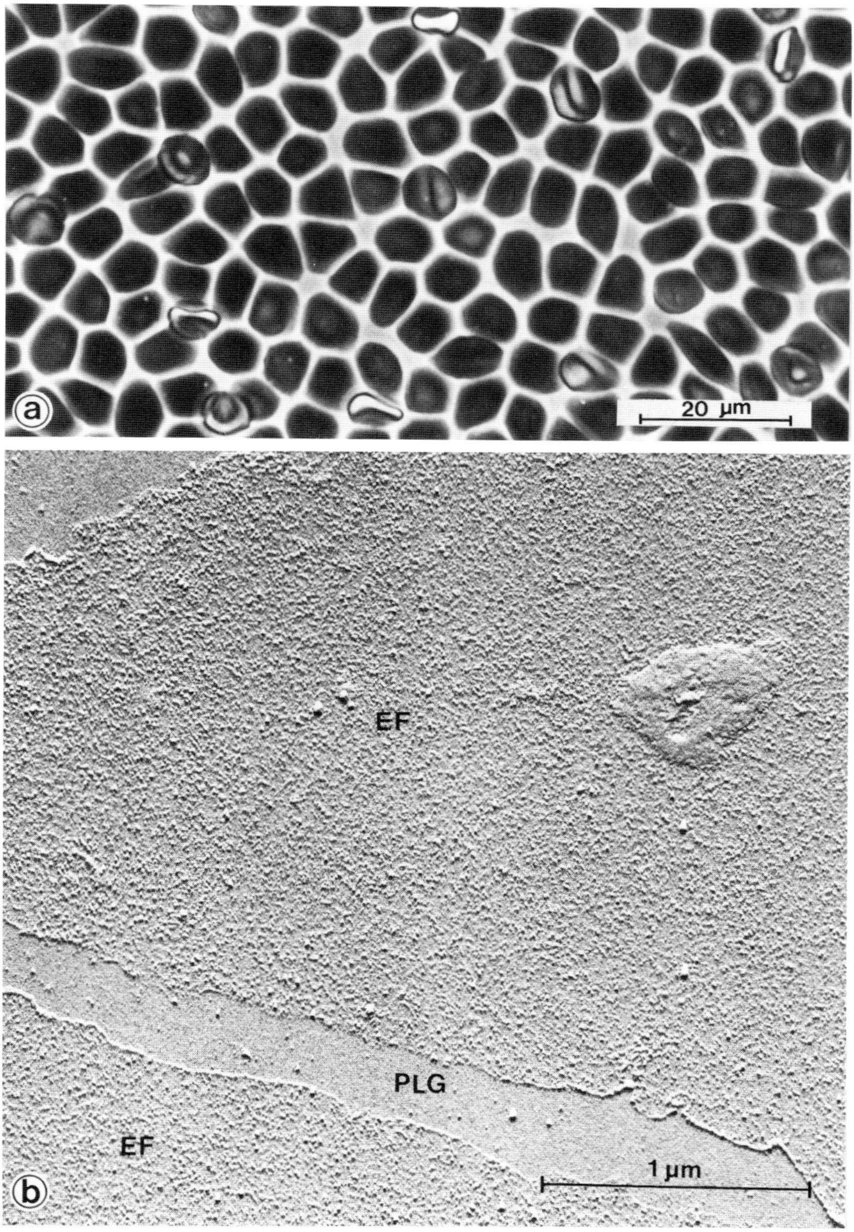

Figure 1. Planar cell monolayers of human erythrocytes. (a) Light micrograph of a monolayer of intact erythrocytes attached to a polylysine-treated glass surface. (b) Electron micrograph of a freeze-fractured erythrocyte monolayer. Split extracellular membrane "halves," E-faces (EF), on polylysine-treated glass (PLG). Platinum carbon shadowing direction for all micrographs, from bottom to top.

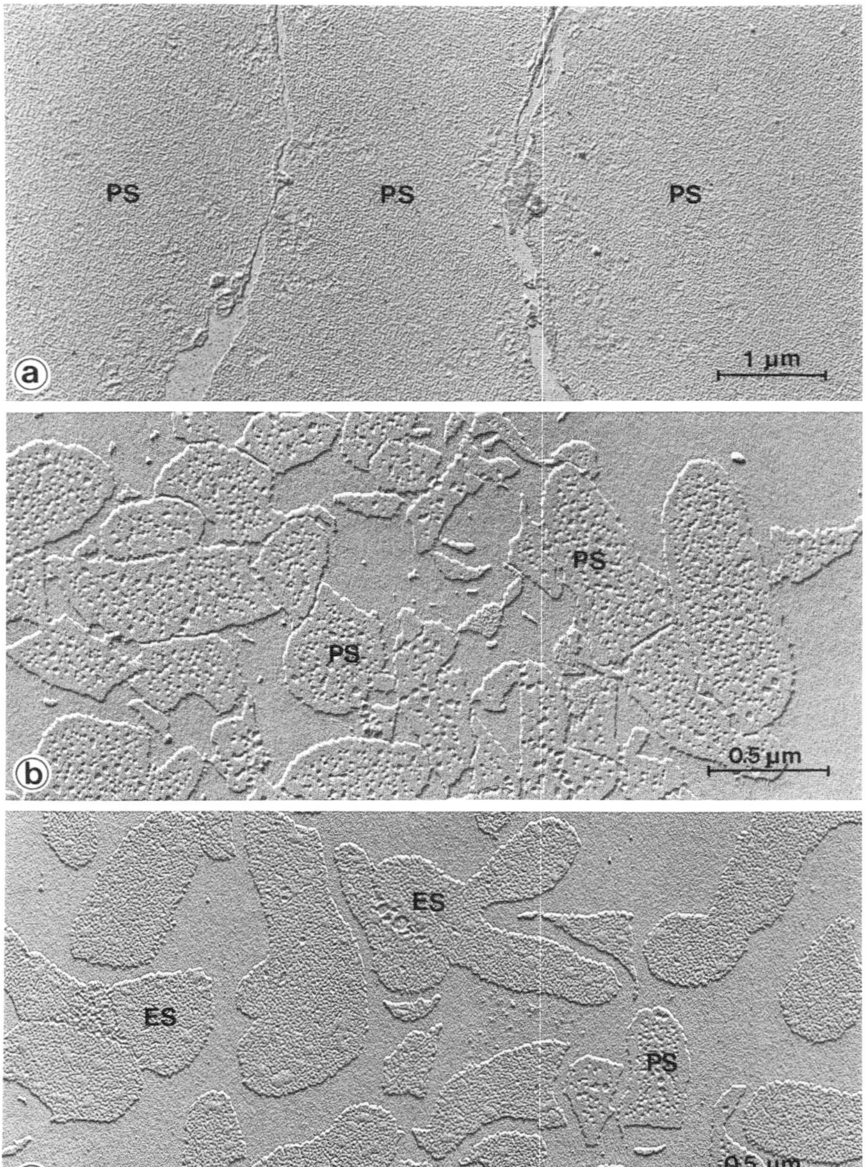

Figure 2. Electron micrographs of single membrane monolayers. (a) Hydraulically sheared human erythrocyte membranes, fixed with glutaraldehyde-osmium, dehydrated with ethanol-acetone, dried with nitrogen gas,

Biochemical studies, especially, require confirmation that hydraulically sheared surfaces retain single planar membranes that are neither overlapped nor folded. Verification can be provided both by light microscopy and by TEM of shadowed freeze-dried planar surfaces. For example, if cell monolayers are "sheared," the removal of cellular contents plus all unattached membrane should expose only the inner cytoplasmic surfaces (PS) of single membranes. Shearing allows chemical or enzymatic modification of an exposed stabilized cytoplasmic surface and provides a means to enrich for the cytoplasmic leaflet of the membrane in membrane splitting studies (discussed below in Section I.E).

D. Monolayer Freeze Fracture of Cells

Cells cultured in Petri dishes often grow to confluence, forming "cell monolayers." Such monolayers are useful for freeze-fracture studies because the fracture plane often passes through the cell as well as along the attached planar plasma membrane (Collins et al., 1975); but they are poorly suited to biochemical studies, which require square centimeter areas of flattened cell membranes. Experimental efforts to flatten membranes by applying them to planar surfaces arose from the observation that membrane bilayers oriented parallel to the plane of fracture split in preference to those of more random orientation (Figure 1b) (Fisher, 1975). Flattened cell membrane monolayers are prepared by applying cells to cleaned planar glass surfaces treated with polylysine. The monolayer is sandwiched against a second glass or copper sheet, then frozen, and fractured. The split preferentially passes through the attached cell membranes.

In biochemical studies planar cell monolayer splitting is used to evaluate the transmembrane distribution of native or probe molecules. By forming flattened sheets it is possible to enrich for the outer leaflet of the plasma membrane. Whereas initial studies required microscopic methods to quantify splitting, improved surface-labeling methods were soon devised (Figure 3). Labeling the extracellular surface of the red cell with fluorescent lectins made it possible to quantify rapidly the amount of outer surface and thus the fraction of outer leaflet of split membrane (Fisher, 1982a). Similarly the

---

Figure 2—continued

and shadowed with platinum-carbon. (b) Purple membrane fragments attached by their extracellular surfaces. Pitted cytoplasmic surfaces (PS) are exposed. (c) Purple membrane fragments attached by their cytoplasmic surfaces. Cracked extracellular surfaces (ES) are exposed. In this preparation 8% of all membranes are oriented with the PS exposed.

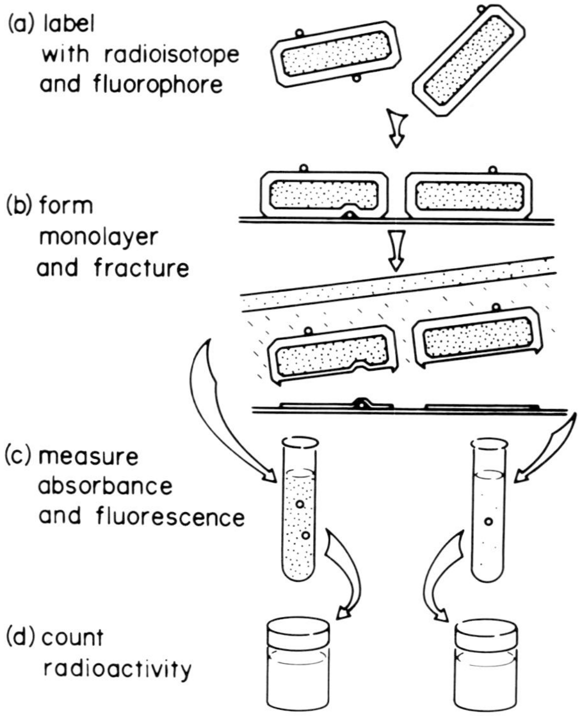

Figure 3. The method of double-labeled membrane splitting (DBLAMS). (a) Washed radio-labeled erythrocytes are surface-labeled with FITC-concanavalin A. (b) Cells are applied to polylysine-treated coverglasses forming a planar cell monolayer that is sandwiched against another piece of glass, frozen, and fractured. (c) The two split fractions are dissolved in buffered detergent solutions, and hemoglobin absorbance and fluorescein fluorescence are measured. (d) Samples are transferred to vials for measurement of radioactivity by LSC. (Reproduced from Fisher, 1982a.)

inner leaflet fraction could also be rapidly quantified by hemoglobin absorbance; the Soret band provides a sensitive spectrophotometric marker of the fraction of cytoplasmic surface. Although the idea of cell monolayer freeze fracture is simple, its application to quantitative biochemical analyses requires careful attention to manipulative detail, avoidance of contamination, and correction for unexpected phenomena such as the enhancement of fluorescence by trace amounts of hemoglobin (Fisher, 1982a).

The quantity of split membrane lipids and proteins produced by fracturing planar surfaces is small. For example, only 5×10^6 red blood cells (RBCs) can bind to an 11×22 mm polylysine-treated coverglass. Given 5.7×10^{-7} mg membrane protein per red cell ghost (Steck, 1974), there is a total of 2.8 μg membrane protein per 11×22 mm glass. Hydraulic shearing further reduces the amount of bound membrane to 50% or less, reducing total membrane protein to 1.4 μg or less. Individual membrane-associated proteins such as actin are thus reduced to nanogram levels, which require highly sensitive methods of detection. Fortunately more sensitive methods continue to be developed, and there are now many techniques to measure nanogram quantities of both lipids and polypeptides.

Cell monolayer freeze fracture has been used to examine the bilayer distribution of both lipids and proteins. For example, in early experiments the transbilayer distribution of native cholesterol was examined using membrane-splitting methods and was verified and quantified by light and electron microscopy (Fisher, 1976a). Native cholesterol in split membrane fractions was measured at the nanogram level by the combination of thin layer chromatography, charring, and microdensitometry. The transbilayer distribution of cholesterol in the intact cell was found to be asymmetric: there was more cholesterol in the outer leaflet than in the inner leaflet of the bilayer (Table 1). The cholesterol study represented the first application of membrane splitting to the biochemical analysis of a molecule of biological interest. The quantitative microscopic approach, although direct, is time consuming, and quantitation requires knowledge or measurement of the total surface area of the cell. Both of these limitations were countered by development of the "doubled-labeled membrane splitting" (DBLAMS) method (Figure 3).

DBLAMS provides a rapid, quantitative biochemical method for analyzing split membrane fractions of whole cell monolayers. It requires having quantifiable markers for each side of the bilayer. This was accomplished for the human erythrocyte by using the fluoresceinated lectin FITC-concanavalin A as a label for the extracellular surface and native hemoglobin as a marker for the inner leaflet (Fisher, 1982a). The quantities of extracellular and cytoplasmic leaflets are thus measured by fluorescence and absorbance spectroscopy. Once the portion of each leaflet in each of the split membrane fractions has been determined, the transbilayer distribution of native and probe molecules can be calculated. For example, in a test study, the transmembrane distribution of radioiodinated concanavalin A was found to be asymmetric. Essentially all of the label was detected on the extracellular surface (ES) of the membrane and partitioned with the extracellular face (EF) upon splitting. In recent studies we have used DBLAMS to determine the transbilayer distribution of tritiated cholesterol. Our initial results suggest that the tritiated probe mimics the distribution of the native molecules and is also asymmetrically distributed. More label is found associated with the extracellular leaflet than with the cytoplasmic leaflet of the bilayer.

Table 1. An Example of the Application of Cell Monolayer Membrane Splitting to an Analysis of the Transbilayer Distribution of Native Cholesterol in Human Erythrocyte Membranes

The ratio of cholesterol in the extracellular leaflet of the bilayer relative to that in the cytoplasmic leaflet (outside/inside) is consistently greater than 1. Thus, cholesterol is asymmetrically distributed across the red cell bilayer, more in the outer leaflet than in the inner leaflet.

	Transparent (%)[b]	*Cholesterol (ng)*[c]			*Ratio*[d]
Exp.[a]	(f_a)	*Cu side* (M_B)	*Glass side* (M_C)	*Total* (M_A)	*Outside/Inside* (m_o/m_i)
PBS 1	88 ± 3 (4)	358 ± 24 (7)	130 ± 25 (7)	488	1.02
2	88 ± 5 (4)	333 ± 60 (3)	143 ± 21 (3)	476	1.64
3	91 ± 8 (4)	350 (1)	125 (1)	475	1.36
TBS 4	87 ± 6 (3)	246 ± 43 (7)	133 ± 13 (7)	379	3.34
5	85 ± 7 (3)	240 ± 37 (8)	128 ± 27 (7)	368	2.39
6	91 (1)	210 ± 32 (4)	86 ± 11 (4)	296	2.01
7	90 (1)	206 ± 8 (4)	96 ± 5 (4)	302	2.75

[a] Exps. 1–3, erythrocytes in 310 imosM phosphate-buffered saline (PBS); Exps. 4–7, in 340 imosM Tris-buffered saline (TBS).

[b] Area determined by planimetry after lipid extraction, drying, and photographic enlargement; mean ± SD. Numbers in parentheses: cover-glasses extracted and planimetered.

[c] Data not vertically comparable: final solvent volume varied among different experiments. Numbers in parentheses; thin-layer chromatography spots examined. Total indicates the arithmetic sum of copper plus glass. In Exps. 6 and 7, control data of bound but unfractured [293 ± 54 (3) and 288 ± 28(3), respectively] and of unbound and unfractured [335 ± 37 (4) and 313 ± 23 (4)] were obtained.

[d] Ratio calculated by applying the analysis given in Fig. 7 using tabular data from f_a, M_A, M_B or M_C, plus $f_{os} = 0.33$ (determined in earlier experiments); m_i and m_o of cholesterol computed for each experiment and their ratio, m_0/m_i, given here. Mean ratio for PBS experiments = 1.34 ± 0.31; for TBS experiments = 2.62 ± 0.57; and for both (overall) = 2.07 ± 0.81.

Source: Reproduced from Fisher (1976a).

In principle, the transbilayer distribution of polypeptides can also be evaluated by DBLAMS. In practice, however, sample preparation procedures for methods such as sodium dodecyl sulfate polyacrylamide gel electrophoresis (SDS PAGE) have interfered with spectrophotometric and fluorometric assays (Fisher and Yanagimoto, 1986). Nevertheless, DBLAMS has shown that freeze fracture of intact erythrocyte membranes does not produce new polypeptide bands; i.e., no new bands can be detected after SDS PAGE by staining with Coomassie Blue or with silver, or by fluorography of lactoperoxidase radioiodinated polypeptides (Figure 4) (Fisher and Yanagimoto, 1986). In other words, freeze fracture does not cleave mem-

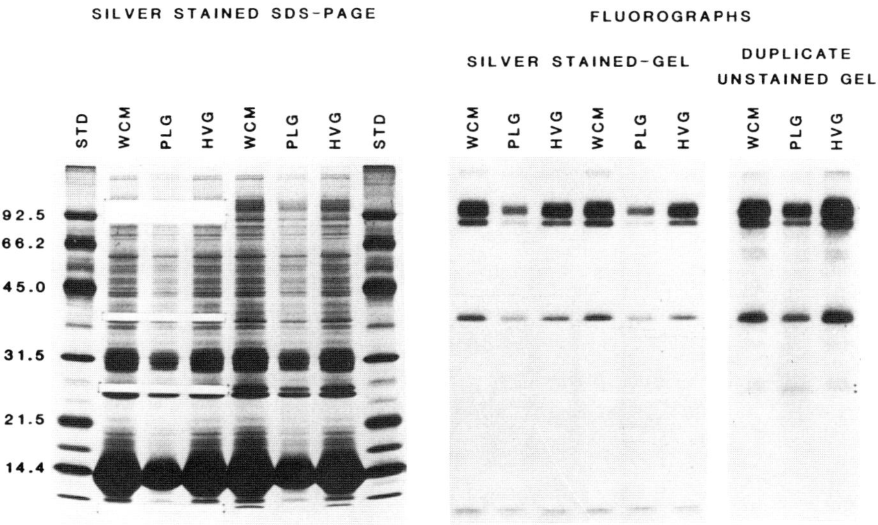

Figure 4. Gel and fluorographs of lactoperoxidase radioiodinated intact and freeze-fractured human red cells. Silver-stained gel (left), its matching fluorograph (center), and a fluorograph of a duplicate unstained gel (right). Gel lanes show duplicate sets of unsplit whole cell monolayers (WCM), split complementary fractions enriched in extracellular (PLG) and cytoplasmic (HVG) membrane "halves" and molecular weight standards (STD) indicated in kilodaltons. Radioactive regions of the left gel were excised for LSC with the center fluorograph used as a template. Note that all bands in the split fractions (PLG and HVG) have matching counterparts in the unsplit control (WCM) whether unlabeled or labeled. Radiolabeled bands include, from top to bottom: band 3 (the anion channel); glycophorin A dimer (PAS 1, just below band 3) and monomer (PAS 2). (Reproduced from Fisher and Yanagimoto, 1986.)

brane-spanning proteins of the human red cell. Moreover, covalent bonds remain intact regardless of the degree of glycosylation. Neither highly glycosylated molecules like glycophorin A, more than 60% of whose mass is composed of carbohydrate (Furthmayr, 1981), nor less glycosylated molecules like the anion channel band 3 are fragmented. Importantly, in planar cell monolayer freeze-fracture studies intrinsic cell–membrane interactions are preserved until splitting. These studies also have a significant weakness, however: only a small percentage of the polypeptides in the split fractions is derived from split membrane. There is a high background of unsplit membrane. As discussed below, this problem was circumvented by the development of SMM splitting methods.

E. Monolayer Freeze Fracture of Single Membranes

Monolayer freeze-fracture of SMM represents a simple but important technical advance. The method involves hydraulically shearing a planar cell, ghost, or vesicle monolayer to form single flattened membranes (Figure 5), as was discussed in Section II.C. Single membrane monolayers are useful for a variety of studies, given certain testable assumptions. First, it is assumed that the topographies of membranes attached to cationic planar surfaces are not altered by lysis and hydraulic shearing. This assumption certainly appears to be valid for membrane polypeptides. For example, comparison by electrophoresis of conventionally prepared red blood cell "ghost" membranes and membranes prepared by cell adsorption, lysis, and shearing showed they were identical. Lysis and shearing of attached intact red cells, however, produces a higher level of hemoglobin relative to membrane

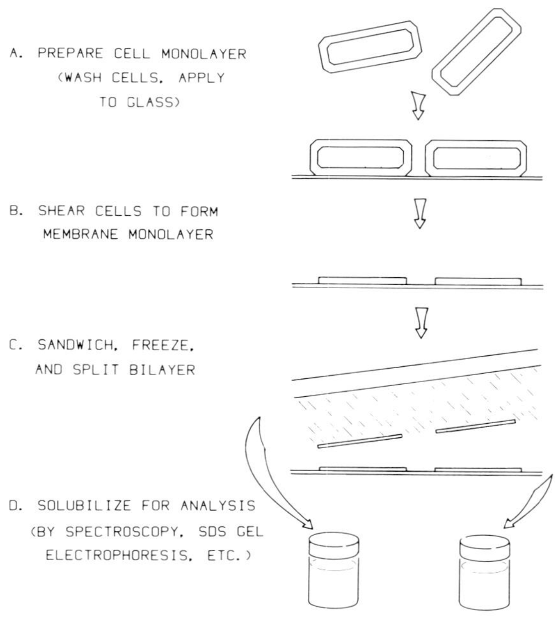

Figure 5. The method of single membrane monolayer splitting (SMMS). (a) Washed cells are applied to polylysine-treated glass. (b) The cell monolayer is lysed, and unbound membrane and cytosol are removed by hydraulic shearing with a forceful jet of buffer. (c) The single membrane monolayer is sandwiched against a second glass, frozen, and fractured. (d) The split fractions are dissolved in buffered detergents for analysis by LSC or SDS PAGE.

proteins than does lysis and shearing of ghosts. Although this is reminiscent of "pink" ghost preparation, some of this hemoglobin is adsorbed to the PLG surface during lysis *in situ.* The planar membrane methods represent a novel approach for rapid isolation of plasma membranes that can be much more precisely controlled than other methods. In addition, the method is useful for electron microscopic studies of the cytoplasmic surface of the membrane (see also Section II.C).

The research in my laboratory has focused on quantitative biophysical and biochemical procedures. For single membrane monolayer splitting (SMMS) studies, our experiments have been almost exclusively limited to two membrane systems: the prokaryotic purple membrane fraction isolated from *Halobacterium halobium* and the plasma membrane of the human erythrocyte. Because these membranes have been well studied structurally and functionally, they are useful for evaluating the effect of freeze fracture on plasma membrane molecules. In addition they represent two structural extremes. Purple membranes are nonvesicular; they are composed of protein and lipid in a weight ratio of about 75% protein to 25% lipid; and single, nonglycosylated, membrane-spanning molecules of the protein bacteriorhodopsin (bR) are packed in a paracrystalline array. In contrast the human red cell membrane is vesicular; it is composed of protein and lipid in a weight ratio of about 50% protein to 50% lipid; the proteins are not in arrays; and transbilayer polypeptides are highly glycosylated.

Purple membranes are nonvesicular sheets that often aggregate in suspension. We found that the preparation of planar monolayers of purple membrane required sonication to remove overlapped or folded membranes (Fisher, 1981). In addition the preparation of oriented planar membrane monolayers required adjusting the pH to titrate the two charged surfaces of purple membrane (Fisher et al., 1977) before application to the cationic glass surface (Figure 2b,c). Planar freeze fracture could then be used to verify the orientation of the membrane as well as to examine the effect of freeze fracture on bR.

Bacteriorhodopsin is an integral membrane polypeptide composed of seven helices that span the lipid bilayer (Unwin and Henderson, 1975; Stoeckenius and Bogomolni, 1981). In our SMMS studies of purple membrane we analyzed split samples by SDS PAGE, detected no new bands, and concluded that freeze fracture did not cleave any covalent bonds (Fisher and Stoeckenius, 1977). Because of the paracrystallinity of bR in purple membrane, however, it seemed risky to speculate that membrane-spanning proteins in general would be unaffected by the splitting process. Indeed, other researchers, using the planar membrane-splitting methods to examine red blood cells, concluded that freeze fracture cleaves covalent bonds of membrane-spanning polypeptides (Edwards et al., 1979; Morrison et al., 1981). In contrast I found that covalent bonds of red cell polypeptides were not cleaved (Fisher, 1982a, 1983). Nermut (1984) also reported that "new bands (indicative of fragmentation of IMP) appear inconsistently." Nevertheless, the idea of covalent bond breakage has been incorporated in a popular cell biology textbook (Alberts et al., 1983).

Recently we have also used SMMS and quantitative methods to show that covalent bonds of RBC band 3 and glycophorin A are not broken (Fisher and Yanagimoto, 1986). Band 3 represents one class of membrane-spanning proteins (Steck, 1978). It is glycosylated on its extracellular surface, contains a 42-kd (kilodalton) cytoplasmic hydrophilic domain that interacts with a variety of cytoplasmic and peripheral membrane proteins such as hemoglobin, ankyrin, and glyceraldehyde-6-phosphodehydrogenase. Band 3 spans the membrane bilayer several times (Ramjeesingh et al., 1984) and may form dimers or tetramers. In contrast, glycophorin A represents another class of membrane-spanning proteins (Furthmayr, 1981). It is heavily glycosylated; more than 60% of its mass is composed of oligosaccharide residues attached to the extracellular polypeptide domain. It appears to interact minimally with cytoplasmic or peripheral membrane proteins, crosses the bilayer only once, and may also interact with itself or with band 3 (Marchesi, 1979). In SMMS studies we labeled these two proteins by lactoperoxidase radioiodination as described by Morrison (1980). Cells were bound to cationic glass, hydraulically sheared, split, quantitatively recovered, and analyzed by SDS PAGE and fluorography, and the radioactivity of the split membrane fractions was measured by liquid scintillation counting (LSC). No new bands could be detected by fluorography, silver staining, or LSC (Figure 6). Thus, for a wide variety of membrane polypeptides, glycosylated or not, spanning the bilayer once or many times, free or multimeric, or randomly associated or arranged in paracrystalline arrays, we have no evidence for covalent bond breakage.

The biochemical observation that membrane-spanning proteins are not cleaved is consistent with cytochemical studies where carbohydrates on split membrane fracture faces are labeled after freeze fracturing (Pinto da Silva and Torrisi, 1982). If covalent bonds were broken, the peptide fragments would probably remain associated with their respective "half" membranes. For example, lectin binding oligosaccharides with disrupted polypeptide chains would be expected to remain with the extracellular leaflet. Instead, their detection on the cytoplasmic leaflet shows that they are pulled through the bilayer. However, a caveat to the argument that the cytochemical studies support the biochemical studies is that the cytochemical methods (e.g., "fracture label") require extensive chemical and physical treatment of the membrane whereas planar membrane freeze-fracture does not. Nevertheless the data are surprisingly similar and mutually compatible. Perhaps differences between the biochemical and cytochemical approaches will be found in the degree of partitioning of proteins with one side of the split membrane

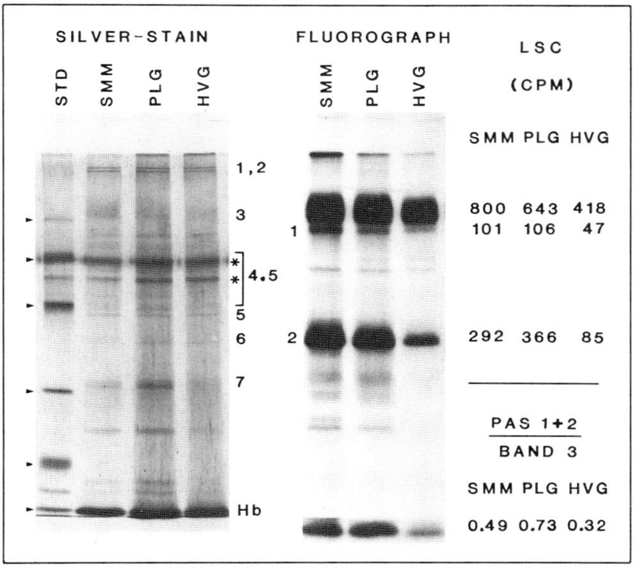

Figure 6. Silver-stained gel (left), complementary fluorograph (center), and quantification of band 3 and glycophorin A (right) of radioiodinated sheared and split single membrane monolayers (SMMS) of human erythrocytes. As in Figure 4, fluorographs were used as templates to excise band 3, PAS 1 (1), and PAS 2 (2) for LSC. Note the presence of both band 3 and glycophorin A on the cytoplasmic (HVG) side and the lack of any new bands. The polylysine-treated glass side (PLG) is enriched in extracellular leaflets, and the heavy-glass side (HVG) is enriched in cytoplasmic leaflets. Also shown are single (unsplit) membrane monolayers (SMM) and molecular weight standards (STD), top to bottom (left arrows): 92.5, 66.2, 45.0, 31.5, 21.5, and 14.4 kd; asterisks indicate regions of spurious bands seen by silver staining. Note the ratios of glycophorin radioactivity (PAS 1 + 2) relative to band 3 for the three fractions SMM, PLG, and HVG (bottom right). There is enrichment of glycophorin over band 3 in the PLG (EF) fraction, and complementary depletion in the HVG (PF) fraction. In other words, glycophorin preferentially partitions with the outer leaflet of the bilayer.

or the other. For example, glutaraldehyde cross-linking of peripheral and integral polypeptides might effect preferential partitioning of the membrane-spanning proteins to the cytoplasmic surface. Alternatively, the cationic interactions of polypeptides with negatively charged extracellular membrane surface sites might cause highly electronegative molecules to partition preferentially to the extracellular surface. This latter possibility seems unlikely, however; the positive charge density of PLG is less than a few percent of the total positive charge bound to its surface (unpublished observations). It should also be mentioned that both biochemical and cytochemical data are in keeping with recent theoretical calculations of the effect of splitting on membrane polypeptides (Keil et al., 1985).

F. Monolayer Freeze-Fracture Autoradiography

Monolayer freeze-fracture autoradiography, or MONOFARG, is a cytochemical method based on planar cell and membrane freeze fracture for detecting the transmembrane and in-plane distribution of isotopic probes (Fisher, 1982a) (Figure 7). Membranes are labeled with radioisotopes of energies compatible with light and electron microscopic autoradiography. Both tritium and iodine-125 have energy levels low enough to ensure high resolution. The technical advances necessary to make monolayer freeze-fracture reproducible and useful have been described (Fisher, 1982a). These included special methods and devices for stabilizing the photographic emulsions using Parlodion films, for applying the stabilized emulsions to samples at low temperatures, and for producing a vacuum-formed fit between replica and emulsion (Fisher, 1982a, 1984).

Given certain assumptions, MONOFARG can be used to determine the transbilayer distribution of extracellular or diffusible ions or molecules such as labeled lipids and hydrophobic or amphipathic probes. It is assumed that the label partitions into the bilayer and resides on one or both sides of the bilayer. It is also assumed that only a small fraction of the label interacts with membrane-spanning proteins, so that lipid partitioning patterns are independent of protein partitioning patterns. It is further assumed that the process of membrane adsorption, flattening, and freezing has little effect on the transbilayer concentration of the probe molecule. Finally, as in any probe study, it is assumed that the distribution of the probe accurately reflects the distribution of the native molecule. Fortunately, each of these assumptions is testable or controllable. It is then a simple matter to compare the amount of radioisotope associated with an intact single membrane to that associated with its split "half" membrane counterpart. For example, if one detected 10 silver grains per square micrometer of intact membrane and 10 silver grains per square micrometer of split EF, one could conclude that 100% of the label remained in the extracellular leaflet; or if there were 10 grains per square micrometer of intact membrane vs. 5 grains per square micrometer of split, one could conclude that the isotope was randomly distributed across the plane of the bilayer, 50% of the isotope partitioning with each complementary half upon splitting.

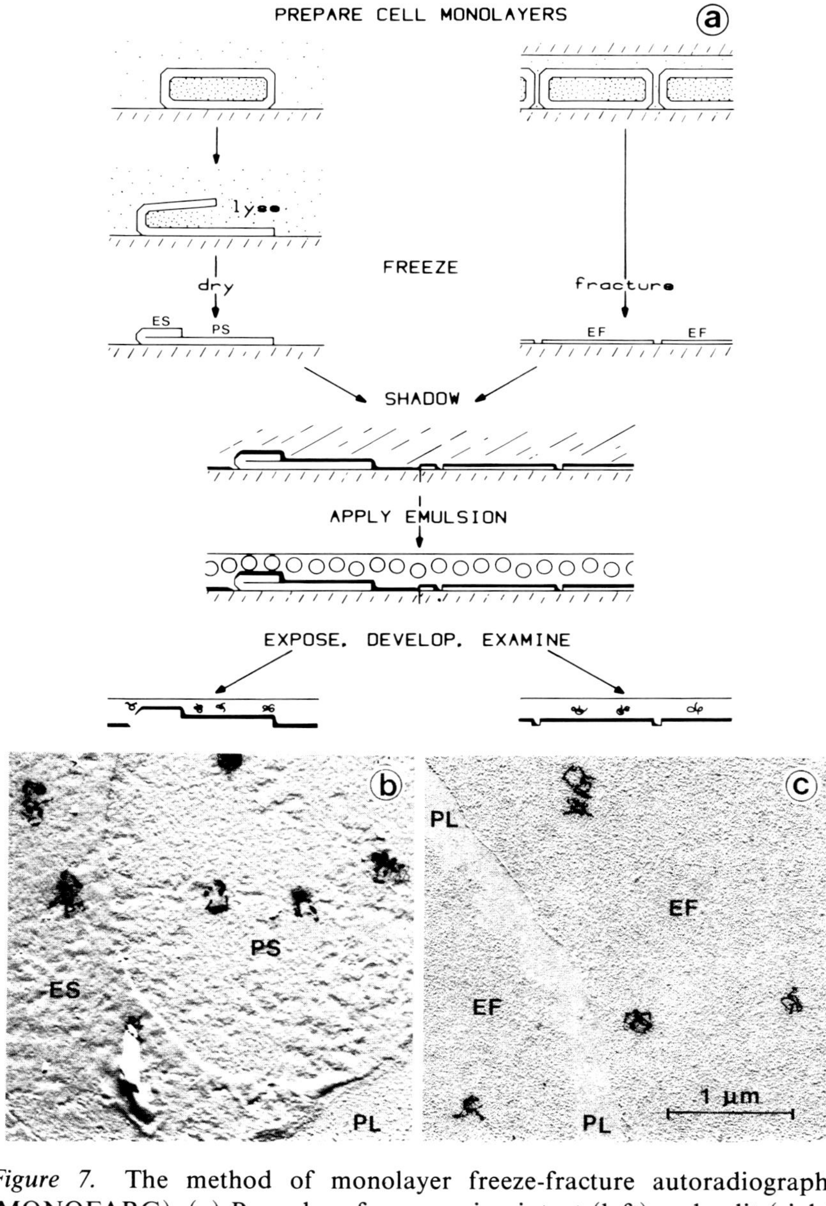

Figure 7. The method of monolayer freeze-fracture autoradiography (MONOFARG). (a) Procedure for preparing intact (left) and split (right) membrane for shadowing, emulsion application, and exposure. (b, c) Electron microscope autoradiographs of intact (b) and split (c) radioiodinated red cell membranes. Cell and split membrane monolayers are prepared as discussed in Figures 3 and 5, then dried and shadowed. Dry stabilized photographic emulsion is applied to the replicas at room temperature, and autoradiographs are exposed, developed, and examined by electron microscopy. (Reproduced from Fisher, 1982b.)

Although resolution or autoradiography is lower than that of gold or ferritin labeling, MONOFARG has the unique ability to evaluate the in-plane distribution of hydrophobic probes or extractable molecules. Such molecules can be stabilized by freezing and, after freeze-drying or "deep etching," detected by application of dry photographic emulsions. There have been diverse applications of the MONOFARG method: MONOFARG, like DBLAMS, has been used to study the transbilayer distribution of tritiated cholesterol (Fisher, 1976b). Under equilibrium labeling conditions, the tritiated label was found to be asymmetrically distributed, at a ratio of 2:1, outside to inside. MONOFARG has also been tested with model systems to evaluate the effect of the replica on resolution and efficiency (Fisher and Banders, 1983).

Initially MONOFARG was used to evaluate the sidedness of concanavalin A after membrane splitting; we found that the lectin remained exclusively with the extracellular surface (Fisher, 1982a). This finding was important for DBLAMS studies since it validated the assumption that FITC–concanavalin A could be used as a quantitative marker for the amount of ES after fracturing as well as before. Our more recent applications of MONOFARG include quantitative studies of lactoperoxidase-labeled erythrocytes (unpublished observations). Cell monolayers were sheared, split, and examined by SDS PAGE and fluorography. Analysis of distribution of radioactivity suggested a highly asymmetric distribution of glycophorin A. For example, results suggested that 84% of the glycophorin polypeptides partitioned to the extracellular side of the split membrane. Knowing the fraction of radioactivity in each radiolabeled polypeptide fraction [e.g., band 3 and PAS bands 1, 2, and 3 (PAS = periodic Schiff's reagent reaction for carbohydrates)] enabled us to evaluate the sidedness of individual polypeptides. Thus it was possible to verify by MONOFARG that over 80% of the radiolabeled glycophorin A partitioned to the ES upon splitting.

III. IMPORTANCE TO FIELD

The methods of planar monolayer freeze fracture and autoradiography of cells and membranes are of relevance to the field of molecular cell biology in several ways. Their significance extends from the level of understanding and formulating first principles of membrane structure to the level of investigating molecular mechanisms of transmembrane signaling.

A. Understanding the Freeze-Fracture Image

Freeze fracture has been an exceptionally important physical tool in studying membranes and in validating a general, widely accepted model for their structure (Singer and Nicolson, 1972). The smooth regions of split membranes are thought to represent the ends of the fatty acyl chains of the lipid bilayer. The IMPs are thought to represent membrane-spanning proteins. But what really is the composition of an IMP? This question was asked in the earliest days of freeze fracture when IMPs were thought to be contaminants—artifacts of the freeze-fracture method. Later indirect evidence suggested that they represented membrane-associated proteins or protein-lipid complexes (Branton and Deamer, 1972). The presence and number of IMPs often directly correlated with protein to lipid ratios of membranes: the more IMPs, the higher the ratio. In addition, membrane reconstitution experiments clearly demonstrated the absence of IMPs in pure lipid systems and their presence in protein plus lipid systems. In fact, with one notable exception, the "lipidic particle" (Verkleij et al., 1979), most IMPs are today tacitly assumed to be composed of integral membrane proteins and/or carbohydrate plus bound lipid. The "lipidic particle" can be demonstrated in synthetic lipid dispersions, but its relevance to biological systems is uncertain. Although it is widely believed that IMPs represent proteins, specifically integral membrane-spanning polypeptides, there has been a dearth of direct chemical information about their composition. However, over the past five years, evidence has begun to accumulate at both biochemical and cytochemical levels.

Planar membrane splitting methods have played an important roll in establishing the biochemical properties of the IMP. For example, using planar splitting one can ask whether the covalent backbone of a transmembrane polypeptide is broken during bilayer splitting. Interestingly, early studies generated conflicting data. In our studies of chemically unmodified purple membrane, we found splitting had no effect on the membrane-spanning polypeptides of bR. However, in qualitative studies of chemically modified red blood cell membranes, other investigators suggested that the polypeptide backbones of membrane-spanning glycoproteins were cleaved. These conclusions were based on data acquired using radiolabeled membranes in combination with planar freeze fracture and SDS PAGE fluorography. Recently, using quantitative approaches we reexamined the effect of bilayer splitting on radioiodinated erythrocyte membranes (Fisher and Yanagimoto, 1986) and single membrane monolayers (Fisher and Yanagimoto, 1985). Our studies suggest that if cleavage of membrane-spanning polypeptides such as glycophorin A or the anion channel does occur, it is quantitatively trivial. Our data strongly suggest that in chemically unfixed, uncryoprotected cell membranes, IMPs are composed of intact polypeptide chains that are often glycosylated (Figure 8).

Given that the IMP is composed of intact polypeptides, one can ask how many polypeptides are in a particle. This question has traditionally been answered using reconstituted systems. The number of particles per unit area of lipid bilayer are counted, and particle number is compared to protein

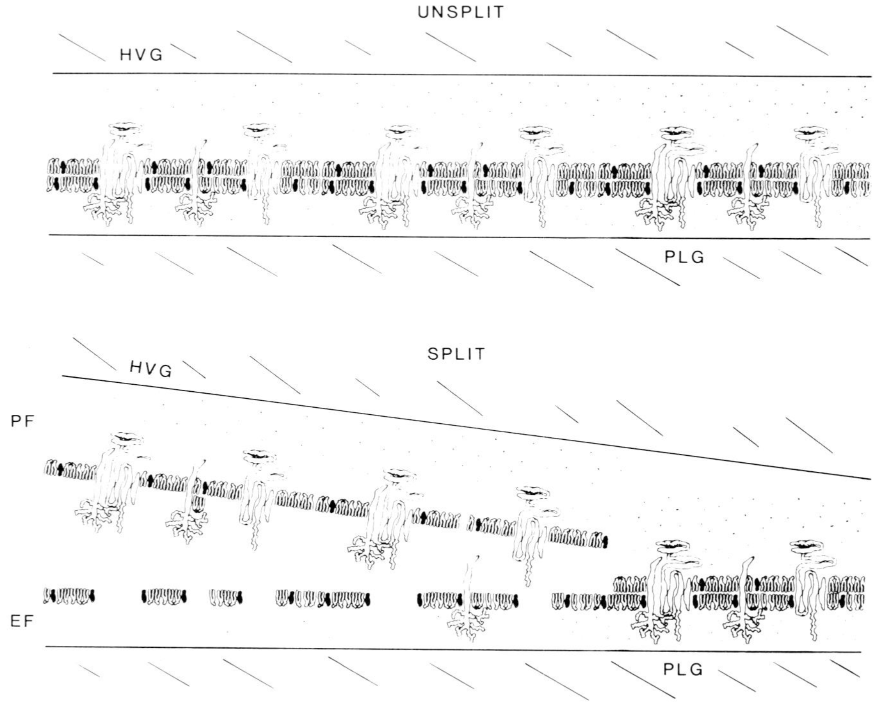

Figure 8. Carbon depicting single membrane monolayer splitting (SMMS) to produce two split red cell membrane fractions. The split PLG side is enriched in extracellular fracture faces (EF), but also contains intact unfractured membrane. The complementary split side (HVG) is significantly enriched in cytoplasmic fracture faces (PF). Two transmembrane proteins (band 3 and glycophorin A) are shown. Band 3 crosses the bilayer several times and possesses a significant cytoplasmic (PS) domain. Glycophorin A is highly glycosylated on the extracellular surface (ES) and spans the bilayer once. Both molecules can interact with themselves and with each other. The cartoon depicts the selective partitioning of band 3 with the cytoplasmic leaflet (PF) and of glycophorin A with the extracellular leaflet (EF). *P*-face IMPs are thus composed of intact, glycosylated polypeptides.

and lipid stoichiometries. In such determinations it is usually necessary to assume that all protein gives rise to IMPs without knowing what fraction may be surface adsorbed, denatured, unincorporated, or partially incorporated. Planar membrane splitting has also been used to answer this question (Fisher and Stoeckenius, 1977). Isolated purple membranes were attached to a planar surface, freeze fractured, shadowed, examined by TEM, and micrographs analyzed by optical diffractometry. A comparison of measured low-resolution lattice spacings to known high-resolution unit cell parameters (Unwin and Henderson, 1975) revealed that each 11-nm IMP was composed of 9–12 intact bR molecules, each of which contained seven membrane-spanning helices: a total of 84 transmembrane helices per particle.

Planar freeze fracture can also provide structural information at the molecular level. For example, purple membrane fragments were oriented by nonrandom adsorption to PLG so that either cytoplasmic or extracellular surfaces were specifically attached. Planar freeze fracture was used to establish which orientation corresponded to attachment of the cytoplasmic surface and which to the extracellular surface. In purple membrane, IMPs partition exclusively to the cytoplasmic side (PF). Similarly oriented purple membranes were examined by electron diffraction (Hayward et al., 1978) to establish that the helices of the bR molecule fan outward on the cytoplasmic side of the membrane.

B. Understanding IMP Partitioning

Although no detailed studies have been published, single membrane monolayers provide an excellent tool for studying the control of the transmembrane distribution of IMPs. For example, one could determine whether the partitioning pattern of membrane-spanning proteins depends on interactions with peripheral or surface proteins. A typical experiment might include preparation of SMM from intact cells to expose cytoplasmic surfaces. Cytoplasmic peripheral proteins could then be totally or selectively modified and the membranes split. Polypeptide partitioning patterns could be analyzed by SDS PAGE and fluorography and quantified by densitometry or LSC. One could also determine whether the hydrophilic domains of the membrane-spanning polypeptides affect their partitioning patterns. The experimenter could treat intact cells with proteases or glycosidases to modify or remove extracellular domains, then form SMMs and use proteases to cleave cytoplasmic domains. Our preliminary experiments with human red cells suggest that the transbilayer distributions of band 3 and glycophorin A after freeze fracture may be influenced by their hydrophilic domains. For example, extraction of all peripheral cytoplasmic proteins by treatment of SMM with 0.1 *N* NaOH appears to have little effect on transbilayer partitioning patterns of glycophorin A and band 3. The question of control of polypeptide partitioning may also be relevant to studies of the molecular details of receptor-effector mechanisms (discussed below in Section III.F).

C. A Novel Approach to Determining Sidedness

It is widely believed that the molecular asymmetry of the membrane has potential functional significance. The importance of single membrane splitting in testing this belief is that this technique allows the asymmetry of the membrane to be studied directly. The structural asymmetry of both lipids and proteins of the red blood cell membrane (Gratzer, 1981) has been well established. Such knowledge, however, has been derived from a collection of often indirect and individually inconclusive methods. Methodological limitations (Carraway, 1975) usually include perturbing the bilayer by chemical modification, e.g., by treatment with lipases or proteases, or comparing the properties of intact cells with those of isolated membranes. Although certain physical approaches, such as electron paramagnetic resonance, nuclear magnetic resonance, and x-ray or neutron diffraction, are noninvasive, they often require studying probes rather than native molecules or require high concentrations of membranes artificially oriented by centrifugation or partial dehydration. Most contemporary cytochemical methods, too, suffer the limitation (or advantage) of chemical modification.

Relative to existing methods for studying membrane structural asymmetries, the membrane-splitting approaches offer several advantages. Chemical or enzymatic modification is not required but is optional. In contrast to other methods for examining membrane sidedness, such as "right-side out" or "inside-out" vesicles (Steck and Kant, 1974), one can easily quantify the amount of protoplasmic membrane surface (PS) exposed and also establish conditions for 100% exposure. Of course a limitation of the SMM preparation is that the membrane is very "leaky": small molecules have access to both sides. Larger molecules and polymers, however, that are physically excluded from the bound surface can be used to examine the exposed PS.

Membranes bound to glass are especially useful for kinetic studies. Coverglasses bearing tenaciously attached membranes can be rapidly and precisely treated with reagents in a stopped-flow type of procedure. Glasses can also be transferred by hand, in seconds, from solution to solution. To detect the regeneration of the bR chromophore, for example, coverglasses with attached, bleached purple membrane were dropped into spectrophotometer cuvettes containing free retinal. Spectral changes could be observed within 200 ms (unpublished observations). In planar-membrane studies native cells or membranes are stabilized by attachment to a cationic surface and, if attached to PLG, to an optically transparent surface. Split "half" membranes are also stabilized and can be examined by light or electron microscopy. A significant advantage to structural studies lies in the fact that the stabilized intact and split membranes are planar. Thus both microscopic and spectrophotometric methods can be used in combination with other biophysical and biochemical approaches.

As with every technique, however, the planar membrane monolayer-splitting methods have limitations. A major limitation for biochemical studies is sample size. As discussed earlier (Section II.D), the amounts of membrane molecules on 11 × 22-mm monolayers lie in the nanogram range. Another potential limitation is the electrostatic interaction between the electropositive glass surface and the electronegative surface of the membrane; this may have an effect at the molecular level. However, the charge density of the PLG surface appears to have little effect on the chemical organization of the adsorbed membrane. Neither lipid nor polypeptide distribution nor transbilayer asymmetry is detectably altered relative to the unattached state. We have also examined the effect of adsorption to PLG on the absorbance peak and absorption maximum of bR (unpublished observations). Visible light spectroscopy of adsorbed purple membranes showed no effect of positive charge on the retinal chromophore, and the spectra of light- and dark-adapted adsorbed membranes were identical to spectra of unbound membranes in suspension. Moreover isolated purple membrane is devoid of a surface layer of carbohydrate, whereas most eukaryotic plasma membranes are glycosylated. One might expect the membrane oligosaccharides to form a protective layer separating the ε-amino groups of the polylysine side-chains from direct interaction with the lipid bilayer of the membrane.

Relative to polypeptide distribution, the sidedness of bulk lipids can be more unambiguously interpreted. The probe or native molecule is expected to partition into one or both of the two leaflets of the bilayer, and freeze fracture is expected simply to separate the two leaflets. This rationale was used in the study of native cholesterol distribution across the plane of the human erythrocyte (Fisher, 1976a). Intact red cell cholesterol was found to be distributed asymmetrically; more was found in the outer leaflet than in the inner leaflet. It should be noted that similar findings have been reported for a fluorescent sterol (Hale and Schroeder, 1982). The transbilayer distribution of cholesterol, however, appears to be quite sensitive to environmental influences and can undergo rapid flip-flop between the two leaflets of the red cell bilayer (Lange et al., 1981). Although the biological relevance of increased cholesterol to phospholipid ratios in the outer leaflet of the cell membrane is unknown, it is tempting to speculate that cholesterol may play a role in making the extracellular surface less susceptible to fusion and/or in altering the permeability properties of the outer leaflet.

Protein distribution after splitting is somewhat more difficult to interpret in terms of sidedness. Membrane-spanning polypeptides, with hydrophilic domains exposed on both sides of the bilayer, can partition to one or the other side of the membrane depending on tertiary or perhaps quaternary interactions. However, in combination with labeling experiments, freeze fracture can demonstrate membrane-spanning functions. For example, a molecule could be labeled exclusively on the extracellular surface. If label partitions to the cytoplasmic side after fracturing, then the molecules either span the membrane or interact with other molecules that functionally span the bilayer. In an experiment to test protein distribution, the sideness of FITC–concanavalin A was quantified using MONOFARG. The concanavalin A was found to be 100% asymmetrically distributed, remaining exclusively with the extracellular leaflet even after membrane splitting. This finding was significant in that it showed that the lectin could be used in DBLAMS experiments as a marker for the fraction of extracellular leaflet and that split membrane fractions could thus be quantified rapidly. Undoubtedly the interactions of peripheral extracellular surface proteins with the bilayer could be similarly examined.

D. Membrane Isolation and Modification

Polycationic beads have been used for rapid isolation of plasma membranes (Jacobson and Branton, 1977). Similarly, polylysine-treated coverglasses can be used to isolate plasma membranes. Monolayers of cells attach to PLG by the extracellular surfaces of their plasma membranes. Hydraulic shearing removes the cytosol and all cytoplasmic membranes, producing a highly purified preparation of plasma membrane. Shearing is undoubtedly one of the most rapid ways to isolate the plasma membrane fraction and has the additional advantage that proteolytic artifacts are minimized during membrane isolation by the exceptionally rapid separation of the cytosolic compartment from the adsorbed membrane. As with membrane isolation using glass beads, however, regions of PLG that lack adsorbed membranes should be pretreated with polyanions, i.e., polycationic blockers, before shearing (Jacobson, 1980).

In biochemical studies of membrane structure and topography, cells and isolated membrane fractions are often chemically modified by exposure to enzymes and reagents. Termination of the reaction often requires physical separation of the chemical participants in the reaction. Although many inhibitors and separation methods have been developed, few have the combination of advantages provided by the adsorbed planar monolayer approach. For example, single coverglasses with bound single membranes can be rapidly washed with—or transferred to—enzyme or inhibitor solutions. I have already mentioned our study on the kinetics of color regeneration of bR "bleached" purple membranes dropped into cuvettes containing free retinal (Section II.E). In another series of experiments the time dependence of proteolysis of the cytoplasmic surface of human erythrocyte membranes was examined. Proteolytic cleavage could be controlled by vigorously washing the PLG surfaces with enzyme or inhibitor solutions, literally starting and stopping the reactions within milliseconds. Precise control of time-dependent processes enables one to ask more detailed questions about the process of enzymatic or chemical modification of the hydrophilic interfaces of the biological membrane. I have mentioned experiments that probe the cytoplasmic surface of the membrane; it is also possible to examine the extracellular surface. For example, given an intact red blood cell monolayer, one can form flattened, collapsed resealed ghost membranes with the unbound extracellular surfaces of the membrane exposed. Nonvesicular membrane sheets such as purple membrane can be selectively oriented before chemical modification.

There is a further advantage to the planar cell or membrane monolayer approach. The structure of the same sample that has been chemically or physically modified can be examined by both light and electron microscopy. Thus structural modifications can be directly compared to biochemical or biophysical observations. In most combined structure and function studies correlations are made by subdividing a sample for microscopy, biochemical assay, etc. With planar monolayers, literally the same cells, membranes, or molecules can be examined by different techniques. For example, in the bR study the same purple membrane sheets that were examined by spectrophotometry were also examined by electron microscopy. Microscopy verified the degree and precision of the planar orientation of the membranes and thus of the bR chromophore relative to the illuminating beam. And microscopy provided precise measurement of the area of membrane, and thus of individual molecules of bR, exposed to the spectrophotometer beam. Few other methods offer the opportunity for such direct comparisons.

E. A Tool for Analyzing Hydrophobic Interactions

An unequivocal forte of the membrane splitting approaches lies with their ability to be applied to questions of hydrophobic interactions. Hydrophobic interactions distinguish membrane molecules from cytoplasmic or water-soluble molecules, and such interactions are notoriously difficult to study by most conventional methods. Freeze fracture, however, is based on the principle that the lipid bilayer—and, by extension, any lipid-lipid or lipid-protein interaction—is weakened by the removal of water. Thus the presence of hydrophobic interfaces can, in principle, be validated or demonstrated by freeze fracture. For example, whereas it might be quite difficult to separate lipid molecules that interact with hydrophobic pockets of membrane-spanning proteins, freeze fracture could conceivably effect such a separation. Although speculative, it should be possible to use planar membrane splitting to evaluate such lipid-protein interactions, especially in model systems or in membranes enriched in a single protein.

A significant advantage for studies of hydrophobic probes or interactions is the paradox of the freezing process: it stabilizes and it destabilizes. It stabilizes or physically fixes the probe *in situ* while concomitantly destabilizing or weakening hydrophobic interactions for subsequent separation by fracturing. Importantly, physical fixation also allows one to examine the distribution of labels or molecules that are soluble or extractable. This should be especially useful in permeability studies.

Three planar cell and membrane monolayer splitting methods have been outlined in this review: DBLAMS, SMMS, and MONOFARG. All can be used to investigate hydrophobic interactions, but each has its own special forte. DBLAMS can be used to examine transbilayer concentrations of native molecules and probes rapidly. Such analysis requires an extracellular surface marker, a cytoplasmic marker, and excellent (>90%) splitting of the monolayer. If markers are unavailable or the percentage of splitting is low, data analysis and interpretation become ambiguous. SMMS can similarly be used to evaluate transbilayer distributions, given appropriate surface markers. For human erythrocyte membrane studies, for example, a variety of cytoplasmic peripheral membrane proteins such as spectrin, actin, and ankyrin can be used to quantify the fraction of cytoplasmic leaflet after splitting. In contrast to DBLAMS, which enriches for the extracellular leaflet, SMMS can enrich for the cytoplasmic leaflet. SMMS applications to date, however, have been limited since the percentage of splitting is often quite low. Moreover, the procedure of lysis and hydraulic shearing may alter the sidedness of particular molecules or probes. Neither DBLAMS nor SMMS can evaluate the differences in distribution of probes among cells or discrete membranes or the distribution of the label within the plane of the intact membrane or split leaflet. However, as a microscopic cytochemical method, MONOFARG has such a capability.

F. Applications to Studies of Signaling Mechanisms

A significant potential application of the freeze-fracture methods described in this review is to studies of the mechanisms of transmembrane signaling. Signaling models have been proposed that invoke interactions of receptor and effector molecules thought to diffuse translationally in the plane of the bilayer. Other models propose diffusion of coupling molecules that may shuttle between receptor and effector. The splitting techniques outlined here could be used to evaluate such interactions. For example, one could imagine potential modulation of receptor-effector interactions as a function of ligand binding. A diffusible receptor interacting with a coupling or effector molecule might form a ligand-induced complex of receptors plus coupling factor and/or effector. An examination of the individual components of the complex by membrane splitting could reveal whether or not the transmembrane distribution of components is altered by ligand binding. Since we have shown that the process of membrane splitting does not cleave the covalent backbones of diverse membrane spanning polypeptides, results could be interpreted in terms of intact protein-protein interactions. Although such expectations are blatantly speculative, they are eminently testable and auger for an exciting future for the planar freeze-fracture approaches.

In conclusion, planar membrane splitting studies lie at a literal and figurative interface: methodologically, at the interface of the disciplines of biophysics, biochemistry, and cell biology; thermodynamically, at the hydrophilic–hydrophobic interface; structurally, at the interface of order vs. disorder; and functionally, at the interface between life and nonlife.

ACKNOWLEDGMENTS

I thank Kathleen C. Yanagimoto for technical help, Eleanor G. Crump for both technical and editorial help, and Karen Schwelm for secretarial assistance. Research cited in this review was supported by National Institutes of Health Grants GM 30983 and GM 31517.

NOTE

1. *Abbreviations and acronyms used*: bR, bacteriorhodopsin; DBLAMS, double-labeled membrane splitting; EF, extracellular fracture-face (extracellular leaflet or "half" membrane); ES, extracellular membrane surface; FITC, fluorescein isothiocyanate; HVG, heavy glass fraction (cytoplasmic leaflet or PF fraction); IMPs, intramembranous particles; LSC, liquid scintillation counting; MONOFARG, monolayer freeze-fracture autoradiography; PAS, periodic acid Schiff's reagent reaction for carbohydrates; PF, protoplasmic fracture-face (cytoplasmic leaflet or "half" membrane); PLG, poly-L-lysine-treated glass; PS, protoplasmic membrane surface; RBCs, red blood cells; SDS PAGE, sodium dodecyl sulfate polyacrylamide gel electrophoresis; SMM, single membrane monolayer; SMMS, single membrane monolayer splitting; STD, standards; TEM, transmission electron microscopy; WCM, whole cell monolayer. Also, kd is the abbreviation for *kilodalton.*

REFERENCES

Aggeler, J., and Werb, Z. (1982) J. Cell Biol. *94:* 613–623.

Alberts, B., Bray, D., Lewis, J., Raff, M., Roberts, K., and Watson, J. D. (1983) *Molecular Biology of the Cell,* p. 273 and Fig. 6-31. Garland, New York.

Alpin, J. D., and Hughes, R. C. (1981) Anal. Biochem. *113:* 144–148.

Bächi, T., and Büechi, M. (1981) in *The Replication of Negative Strand Viruses* (D. H. L. Bishop and R. W. Compans, eds.), pp. 553–558. Elsevier, Amsterdam and New York.

Bearer, E. L., and Friend, D. S. (1980) Proc. Natl. Acad. Sci. USA *77:* 6601–6605.

Branton, D. (1966) Proc. Natl. Acad. Sci. US *55:* 1048–1056.

Branton, D., and Deamer, D. W. (1972) *Protoplasmatologia,* Vol. II.E.1, 70 pp. Springer-Verlag, New York.

Büechi, M., and Bächi, T. (1979) J. Cell Biol. *83:* 338–347.

Carraway, K. L. (1975) Biochim. Biophys. Acta *415:* 379–410.

Clarke, M., Schatten, G., Mazia, D., and Spudich, J. A. (1975) Proc. Natl. Acad. Sci. USA *72:* 1758–1762.

Collins, T. R., Bartholomew, J. C., and Calvin, M. (1975) J. Cell Biol. *67:* 904-911.
Edwards, H. H., Mueller, T. J., and Morrison, M. (1979) Science (Wash. DC) *203:* 1343-1345.
Elias, P. M., Friend, D. S., and Goerke, J. (1979) J. Histochem. Cytochem. *27:* 1247-1260.
Fisher, K. A. (1975) Science (Wash. DC) *190:* 983-985.
Fisher, K. A. (1976a) Proc. Natl. Acad. Sci. USA *73:* 173-177.
Fisher, K. A. (1976b) J. Cell Biol. *70:* 218a.
Fisher, K. A. (1978) in *Electron Microscopy* (W. M. Sturgess, ed.), pp. 521-532. Microscopical Society of Canada, Toronto.
Fisher, K. A. (1980) Annu. Rev. Physiol. *42:* 261-273.
Fisher, K. A. (1981) Methods Enzymol. *88:* 230-235.
Fisher, K. A. (1982a) J. Cell Biol. *92:* 44-52.
Fisher, K. A. (1982b) J. Microsc. (Oxf.) *126:* 1-8.
Fisher, K. A. (1983) Biophys. J. *41:* 118a.
Fisher, K. A. (1984) in *The Science of Biological Specimen Preparation* (J. P. Revel, T. Bernard, and G. H. Haggis, eds.), pp. 189-194. SEM Inc., AMF O'Hare, IL.
Fisher, K. A., and Banders, U. T. (1983) J. Cell Biol. *97:* 308a.
Fisher, K. A., and Stoeckenius, W. (1977) Science (Wash. DC) *197:* 72-74.
Fisher, K. A., and Yanagimoto, K. C. (1985) J. Cell Biol. *101:* 307a.
Fisher, K. A., and Yanagimoto, K. C. (1986) J. Cell Biol. *102:* 551-559.
Fisher, K. A., Yanagimoto, K., and Stoeckenius, W. (1977) J. Cell Biol. *75:* 220a.
Fisher, K. A., Yanagimoto, K., and Stoeckenius, W. (1978) J. Cell Biol. *77:* 611-621.
Furthmayr, H. (1981) in *Biology of Carbohydrates* (V. Ginsburg and P. Robbins, eds.), pp. 123-198. Wiley, New York.
Gratzer, W. B. (1981) Biochem. J. *198:* 1-8.
Hale, J. E., and Schroeder, F. (1982) Eur. J. Biochem. *122:* 649-661.
Hayward, S. B., Grano, D. A., Glaeser, R. M., and Fisher, K. A. (1978) Proc. Natl. Acad. Sci. USA *75:* 4320-4324.
Jacobson, B. S. (1980) Biochim. Biophys. Acta *600:* 769-780.
Jacobson, B. S., and Branton, D. (1977) Science (Wash. DC) *195:* 302-304.
Kell, M. J., Defelice, L. J., and Goolsby, W. (1985) J. Theor. Biol. *115:* 103-127.
Lang, R. D. A., and Nermut, M. V. (1980) Cell Biol. Int. Rep. *4:* 751.
Lange, Y., Dolde, J., and Steck, T. L. (1981) J. Biol. Chem. *256:* 5321-5323.
Marchesi, V. T. (1979) Semin. Haematol. *16:* 3-20.
Maurer, A., and Mühlethaler, K. (1981) Eur. J. Cell Biol. *24:* 216-225.
Mazia, D., Schatten, G., and Sale, W. (1975) J. Cell Biol. *66:* 198-200.
Morrison, M. (1980) Methods Enzymol. *70:* 214-220.
Morrison, M., Mueller, T. J., and Edwards, H. H. (1981) in *The Function of Red Blood Cells: Erythrocyte Pathobiology* (D. F. H Wallach, ed.), pp. 17-34. New York.
Nermut, M. V. (1981) Proc. R. Microsc. Soc. *16:* 178.
Nermut, M. V. (1982) Eur. J. Cell Biol. *28:* 160-172.
Nermut, M. V. (1983a) Eur. J. Cell Biol. *31:* 296-304.
Nermut, M. V. (1983b) Trends Biochem. Sci. *8:* 303-306.
Nermut, M. V. (1984) in *The Science of Biological Specimen Preparation* (J. P. Revel, T. Barnard, and G. H. Haggis, eds.), pp. 181-187. SEM Inc., AMF O'Hare, IL.
Nermut, M. V., and Williams, L. D. (1977) J. Microsc. *110:* 121-132.
Nevo, A., DeVries, A., and Katchalsky, A. (1955) Biochim. Biophys. Acta *17:* 536-547.
Pinto da Silva, P. (1972) J. Cell Biol. *53:* 777-787.
Pinto da Silva, P., and Branton, D. (1970) J. Cell Biol. *45:* 598-605.
Pinto da Silva, P., and Nicolson, G. L. (1974) Biochim. Biophys. Acta *363:* 311-319.
Pinto da Silva, P., and Torrisi, M. R. (1982) J. Cell Biol. *93:* 463-469.
Pinto da Silva, P., and Kan, F. W. K. (1984) J. Cell Biol. *99;* 1156-1161.
Pinto da Silva, P., Douglas, S., and Branton, D. (1971) Nature (Lond.) *232:* 194-196.

Pinto da Silva, P., Kachar, B., Torrisi, M. R., Brown, C., and Parkison, C. (1981) Science (Wash. DC) *213:* 230–233.

Ramjeesingh, M., Gaarn, A., and Rothstein, A. (1984) Biochim. Biophys. Acta *769:* 381–389.

Rash, J. E., Johnson, T. J. A., Hudson, C. S., Giddings, F. D., Graham, W. F., and Eldefrawi, M. E. (1982) J. Microsc. *128:* 121–138.

Singer, S. J., and Nicolson, G. L. (1972) Science (Wash. DC) *175:* 720–731.

Sommer, J. R. (1977) J. Cell Biol. *75:* 245a.

Steck, T. L. (1974) J. Cell Biol. *62:* 1–19.

Steck, T. L. (1978) J. Supramol. Struct. *8:* 311–324.

Steck, T. L., and Kant, J. A. (1974) Methods Enzymol. *31:* 172–180.

Stoeckenius, W., and Bogomolni, R. (1981) Annu. Rev. Biochem. *52:* 587–616.

Unwin, P. N. T., and Henderson, R. (1975) J. Mol. Biol. *94:* 425–440.

Verkleij, A. J., De Kruijff, B., Gerritsen, W. F., Demel, R. A., Van Deenen, L. L. M., and Ververgaert, P. H. J. (1973) Biochim. Biophys. Acta *291:* 577–581.

Verkleij, A. J., Mombers, C., Leunissen-Bijvelt, J., and Ververgaert, P. H. J. (1979) Nature Lond. *279:* 162–163.

THE BRUSH BORDER:
A MODEL FOR STRUCTURE, BIOCHEMISTRY, MOTILITY, AND ASSEMBLY OF THE CYTOSKELETON

David R. Burgess

I. INTRODUCTION

The absorptive cell of the intestinal epithelium is a highly polarized columnar cell 20 μm in height whose chief function is the transport of nutrients. The absorptive cells, along with the mucus-secreting goblet cells and a few other specific cell types, form a simple epithelium in the shape of ~2 mm in length macroscopic finger-like or paddle-shaped villi. Between the densely packed villi, the epithelium is deeply infolded into crypts where an actively mitotic epithelial cell population that gives rise to the differentiated villus cells exists. The nondividing cells migrate over a 2- to 3-day period along

Advances in Cell Biology, Vol. 1, pages 31–58.

ISBN: 0-89232-792-8

the length of the villus to the tip, where they die and slough off into the intestinal lumen. Nutrient transport is facilitated by the fact that the luminal, or apical, plasma membrane is infolded into more than a thousand finger-like microvilli, 1.5–2 μm in length. Not only is cell shape asymmetric, but the protein distribution of the plasma membrane is polarized with many transport enzymes localized to the apical membrane whose domain remains separate from the basolateral membrane, itself containing unique proteins, by extensive tight junctions (see Danielsen et al., 1984, for review). Within microvilli, and providing their structural support, is a core of 10–50 actin microfilaments that are extensively cross-linked to one another and to the overlying plasma membrane. The bundled actin filaments penetrate about 0.7 μm into the apical cytoplasm, where they terminate at the plane of the apical desmosomes. The bundle of actin and associated proteins within the microvillus proper comprises one domain of the brush border, termed the microvillus core. The portion of the bundle of filaments extending into the apical cytoplasm is called a rootlet. The other domain of the brush border is demarcated by the apical junctional complex composed of the tight, intermediate, and desmosomal junctions (Farquhar and Palade, 1963). The filament systems of the junctional complex, of the rootlets, and that between the rootlets is called the terminal web. Because there has been so much work on the brush-border cytoskeleton during the past 15 years or so it has become a useful model for comparative studies on the organization, biochemistry, and motility of other cytoskeletons. This review will focus on the morphological organization, biochemistry, and development of the brush-border cytoskeleton. Other recent reviews (primarily dealing with structure and biochemistry) have also been published on this interesting system (Mooseker, 1985; Bretscher, 1983a). I will also attempt to delineate areas for future exploration or aspects of the brush-border cytoskeleton that have been ignored and deserve attention. Our current model for the organization of the brush border is depicted in Figure 1, in which every attempt has been made to draw all elements to scale. Table 1 presents a list of brush-border cytoskeletal proteins known to date.

II. THE MICROVILLUS DOMAIN

A. Structure

Our knowledge of the ultrastructure of the microvillus core has advanced since the early 1950s largely due to advances in instrumentation and specimen preparation. The earliest studies revealed the brush border to be composed of parallel, closely packed microvilli (Granger and Baker, 1950; Dalton, 1951). Intestinal brush border microvilli are remarkably constant

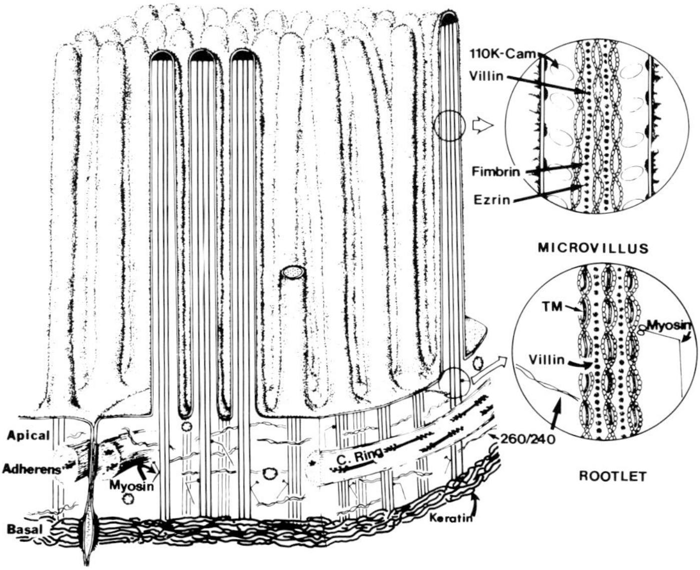

Figure 1. Current model for the organization of the brush border, with all elements drawn to scale.

in length and typical for each organism in the adult, although in general length is greater in the upper one-third of a villus (Palay and Karlin, 1959; Brown, 1962; Burgess and Prum, 1982). Microvillus length is also thought to vary along the extent of the intestine, being longest in the duodenum. Microvilli are sparse and short on undifferentiated crypt cells, and upon reaching the base of the villi the newly nonmitotic cells begin to acquire both ultrastructurally distinct features and specific apical membrane enzymes. Counts of microvillar number per square micrometer on cells of the upper one-third of villi in adult chickens have led to the estimate of 1,200 microvilli per cell (Stidwill and Burgess, 1986).

A bundle of thin filaments running parallel to the long axis of the microvillus and extending into the terminal web is the most distinctive feature of the core and was recognized early (Palay and Karlin, 1959; Millington and Finean, 1962; Overton and Shoup, 1964; Overton, 1965; Overton et al., 1965; Ito, 1965; Mukherjee and Wynn Williams, 1967; Bonneville and Weinstock, 1970; Brunser and Luft, 1970; Mukherjee and Staehelin, 1971). The number of core filaments, ranging between 10 and 50, appears to be variable between species, but constant within one species. Actual counts of the number of core filaments from 10-day-old chicks is 15 (Stidwill and Burgess, 1986), whereas the micrographs published by Brunser

Table 1. Brush Border Cytoskeletal Proteins

Protein	Molecular Weight	Putative Function
A. Microvillar Core		
120K	120	Putative 110K-binding membrane protein
110K-calmodulin[a]	110, 16	Actin-membrane link, Ca^{++}-buffer
Villin[a]	95	(1) Ca^{++}-activated actin severer (2) Secondary actin bundler
Ezrin[a]	81	?
Fimbrin[a]	68	Primary actin bundler
Actin[a]	42	Structural Support
110K-kinase		?
Ezrin kinase		?
B. Terminal Web		
1. Rootlets		
Caldesmon	?	Ca^{++}-regulated actin-calmodulin binding
Villin	95	Actin bundler
Actin	42	Structural support
Tropomyosin	30–32	(1) Stabilize actin (2) Activate myosin
2. Inter-rootlet		
Myosin	200, 19, 16	Putative vesicle movement
TW 260/240	260, 240	Crosslink rootlets for structural support
3. Circumferential ring		
Myosin	200, 19, 16	Contraction of ring
Vinculin	130	Actin-membrane association
α-Actinin	95	Actin-membrane association
Actin	42	Contraction and structure of ring
Tropomyosin	30–32	Stabilize ring and activate myosin
4. Not specifically localized		
Filamin	250	?
Gelsolin	91	?
Calpactin	36, 10	?, putative actin membrane binding

Table 1—continued

Protein	*Molecular Weight*	*Putative Function*
Intestinal Ca^{++}-binding protein	28	?, Ca^{++}-buffer
Calmodulin	16	?, Ca^{++}-buffer
MLCK		Regulating myosin
MLCP		Regulating myosin
Tyrosine kinase		Regulating calpactin

[a] Molar ratio of actin 110K : villin : fimbrin : calmodulin : ezrin is approximately 10 : 1.1 : 1.6 : 1.3 : 4.0 : 0.07.

and Luft (1970) suggest that there are about 50 core filaments in rat intestinal microvilli. Our estimates of the number of core filaments from adult chickens is in the range of 15–20. The filaments themselves are loosely hexagonally packed in arrangement and terminate, or anchor into, electron-dense material at the microvillus tip (Brunser and Luft, 1970; Mukherjee and Staehelin, 1971).

The other major distinctive ultrastructural features of the microvillus core are the periodically spaced cross-bridges that link the core filaments laterally to electron-dense points on the plasma membrane (Millington and Finean, 1962; Mukherjee and Staehelin, 1971). Upon treatment of isolated brush borders with high-Mg^{++} solutions, the periodic nature of the cross filaments became apparent (Mooseker and Tilney, 1975). This 33-nm periodicity of the cross filaments was apparent without high-Mg^{++} treatment in intact cells or in membranated or demembranated isolated microvilli and brush borders (Mukherjee and Staehelin, 1971; Mooseker and Tilney, 1975; McNutt, 1978; Bretscher and Weber, 1978a, 1979; Matsudaira and Burgess, 1979, 1982b). These studies revealed the cross filaments to be 2–7 nm in width and 15–30 nm in length. An analysis of the organization of the cross filaments suggests that they are in fact composed of a double helix of bridges spiraling around the core filament bundle (Matsudaira and Burgess, 1982b). This study also found that the cross filaments are slanted at an angle such that they appear to form an arrowhead complex with the core filaments, with the point of the arrow directed toward the cell center. Some freeze-fracture studies have indicated that the cross filaments are attached to membrane particles (Mukherjee and Staehelin, 1971); however, the membrane particle distribution does not correlate with cross-filament periodicity (Mukherjee and Staehelin, 1971; McNutt, 1978; Neutra, 1979). Interestingly, high-Mg^{++} treatment leads to a periodic helical membrane-particle distribution, but the spacing is 90 nm, which is three times larger than the cross-filament periodicity (Mooseker and Tilney, 1975). At this time we know very little about the structural interaction between cross filaments and the membrane.

B. Proteins

The past 15 years have seen an explosion in knowledge about the structural proteins of the brush border cytoskeleton. This burst in information is largely due to the fact that large numbers of absorptive cells can now be easily isolated, from which both brush borders and microvilli can be purified. Particular credit is due to Miller and Crane (1961) and to Bretscher and Weber (1978a), who described practical methods for the isolation of brush borders and microvilli, respectively. An unfortunate contaminant of the brush-border preparation, which is particularly difficult to eliminate, is deoxyribonucleic acid (DNA) and its associated proteins; this problem will be addressed later. Another advance was the finding by Mooseker (1976) that the brush border remained structurally intact after many of the membrane proteins are solubilized by Triton X-100, yielding the brush-border cytoskeleton. Not only do intact brush borders retain their integrity after Triton treatment, but isolated microvillar cytoskeletons retain their bundled filaments, cross filaments, and dense tips (Bretscher and Weber, 1978a; Matsudaira and Burgess, 1979).

1. Actin

The dominant protein of the microvillus core is also that of the microfilament. As was orginally suggested in the pioneering work of Ishikawa et al. (1969), in which the muscle myosin fragment HMM (heavy meromyosin) was first used to identify actin filaments in nonmuscle cells, and then confirmed by Tilney and Mooseker (1971), the microfilaments are composed of actin. All of the actin filaments within the core are unidirectionally polarized with the HMM arrowhead pointing toward the cell center (Mooseker, 1976; Begg et al., 1978).

The actin is a typical nonmuscle actin composed of a 40:60 ratio of the β and γ isoforms (Bretscher and Weber, 1978a). Since the actin isoforms were purified from whole brush borders, it is not known whether the β or γ isoforms are restricted to different locations within the brush border (i.e., circumferential ring vs. cores), although such a possibility seems unlikely.

2. Actin-Associated Proteins

The microvillus core is particularly rich in actin-binding proteins fulfilling various roles. Two well-characterized proteins, villin and fimbrin, act as the major filament-bundling proteins; a 110-kd protein (hereafter called 110K) complexed with calmodulin makes up the cross filaments; and ezrin is a newly identified actin associated core protein. The breakthrough as to the identity of the major core proteins came with Bretscher and Weber's (1978b, 1979) reports on microvillus isolation and on villin as a core protein. More detail on the composition of the core came from our work with isolated cores in which ATP was found to dissociate cross filaments and remove the 110- and 16.5-kd polypeptides from the core bundle, whereas deoxycholate treatment unbundled the filaments (but left the cross filaments attached) and removed both 95- and 68-kd polypeptides from the core (Matsudaira and Burgess, 1979). While the 95-kd [or villin] protein had been thought to be both the cross-filament protein and a filament-bundling protein, our work suggested that it serves primarily as a bundler of actin. The core is now thought to be composed of five major proteins: actin, 110K, villin (95K), 68K (later named fimbrin), and 16.5K (later identified as calmodulin). The molar ratios of these proteins is approximated to be 10 actin: 1.1 110K : 1.6 villin : 1.3 fimbrin : 4.0 calmodulin (Matsudaira and Burgess, 1979; Glenney and Weber, 1980; Glenney et al., 1980; Howe et al., 1980). The identification of these four major actin-associated proteins as being core proteins has been confirmed by immunolocalization studies, which have also been a powerful tool for the dissection of the terminal web region. Contaminants of microvillar core preparations and brush borders are nuclear proteins in the 12–16-kd range (Glenney and Weber, 1980).

a. Villin. Villin was the first actin-binding protein of the core to be isolated and characterized. It is by far the best-characterized core filament actin-binding protein and has been reviewed in depth recently (Mooseker, 1985). Since its properties have been detailed, I will only summarize them here; I will discuss some of the newer work on villin and cover functional implications of this key protein.

It should be pointed out first that villin is not related to α-actinin. A report that antibodies against α-actinin-labeled microvilli (Schollmeyer et al., 1974) led to the exciting but incorrect suggestion that the microvillus tips and even the cross filaments were composed of α-actinin (Mooseker and Tilney, 1975). The possibility of relatedness between villin and α-actinin persisted until 1980, when careful peptide maps and specific antibodies showed little similarity (Craig and Lancashire, 1980; Mooseker and Stephens, 1980). More importantly, the properties of the two proteins are distinct.

Villin is one member of a large class of actin-binding proteins that show Ca^{++}-dependent actin-binding properties. The initial characterization of villin showed it to be an actin filament-bundling protein at low Ca^{++} concentrations ($<10^{-6}$) and a filament-severing protein at elevated free

Ca^{++} concentrations ($>10^{-6}$) (Bretscher and Weber, 1980b; Mooseker et al., 1980; Craig and Powell, 1980; Matsudaira and Burgess, 1982a). Aside from these basic properties, villin has been shown to have three high-affinity Ca^{++}-binding sites whereby upon binding Ca^{++} the protein becomes more elongated (Hesterberg and Weber, 1983a,b). Detailed studies on the profound effects of villin on actin assembly, actin bundling, and actin severing have been conducted with both intact villin and with villin fragments generated by specific proteases. Villin's effects on actin assembly are only evident above micromolar free Ca^{++} levels and include capping activity at the barbed end of the actin filament and enhancing the nucleation phase, which is the rate-limiting step, of actin assembly (Mooseker et al., 1980; Cribbs et al., 1982; Glenney et al., 1981c; Wang et al., 1983; Walsh et al., 1984a,b). Villin will saturate actin filaments at a molar ratio of about 1:2, villin to actin. Villin's severing activity is blocked by tropomyosin (Mooseker et al., 1982a; Bonder and Mooseker, 1983; Walsh et al., 1984a), a finding rich in func- tional implications, since the rootlets but not the core possess tropomyosin.

The various functional domains of villin have begun to be mapped using site-specific cleavage methods. This methodology was first used by Glenney et al. (1981a) and Glenney and Weber (1981), who showed that villin could be cleaved into a 8.5-kd headpiece, which binds actin Ca^{++} independently, and a 90-kd core, which, while lacking bundling ability, has Ca^{++}-dependent nucleating and severing abilities. More recently, Matsudaira and co-workers (1985a,b) and Hesterberg and Weber (1986) have extended this technique with the use of a fleet of sequential cleavages coupled with region-specific antibodies in an attempt to localize functions to protein domains. While this technique may hold much promise, it will surely be complemented by molecular cloning methodologies. There is a preliminary report of a human villin clone derived from the HT-29 cell line yielding partial sequence information (D. Louvard, personal communication). The protein domain mapping methodology depends on the risky but plausible assumption that fragments will retain the correct tertiary structure and not lose or gain functions.

Since tropomyosin is restricted to the terminal web it is possible that competition with core actin-binding proteins might explain its absence from the core. Therefore we initiated a study on the effects of villin on the binding of intestinal epithelial cell tropomyosin to actin (Broschat et al., 1985; Burgess et al., 1987). In fact, we found that villin, which has an apparent K_a of $6 \times 10^5\ M^{-1}$ for F-actin, is a poor competitor of tropomyosin binding to villin saturated F-actin bundles. On the other hand tropomyosin, with an apparent K_a of $8 \times 10^6\ M^{-1}$ for F-actin, inhibited full saturation of villin to tropomyosin saturated filaments. Somehow, tropomyosin is restricted to the rootlet. It is likely that another core protein or the core proteins as a whole inhibit tropomyosin's binding to the core. This interesting issue has yet to be resolved.

b. Fimbrin. Fimbrin belongs to another class of actin-binding proteins whose properties are independent of Ca^{++}. Fimbrin is a potent filament-bundling protein; however, as tested, its optimal bundling abilities occur at unphysiologically low Mg^{++} and salt concentrations (Bretscher, 1981; Glenney et al., 1981b). Fimbrin will saturate F-actin at a molar ratio of 1:3, fimbrin to actin (Bretscher, 1981; Glenney et al., 1981c), a value similar to that of villin's. Fimbrin appears to possess one divalent cation-binding site with a stronger affinity for Mg^{++} than for Ca^{++}, indicating that the Ca^{++} site is not of physiological importance (Glenney et al., 1981b). Interestingly, fimbrin and villin when co-mixed with F-actin will enhance each other's binding to actin (Glenney et al., 1981b). Actin bundles formed with fimbrin are more ordered than those formed with villin (Matsudaira et al., 1983) and are unidirectionally polarized, suggesting that fimbrin may be the primary bundler of core filaments (Glenney et al., 1981b). Bundles formed by a mixture of fimbrin and villin, at approximate physiological levels, are also unidirectionally polarized (Glenney et al., 1981b). Fimbrin, or a very closely related actin-binding protein, is widely distributed, as it has been localized by immunofluorescence to ruffled membranes, microspikes, stereocilia, and microvilli of various tissue culture cells (Bretscher and Weber, 1980a).

c. The 110K-calmodulin complex. One of the more intriguing components of the core is the 110K-calmodulin complex thought to make up the cross filaments. Glenney et al. (1982c) have used antibodies prepared against 110K in immunoferritin localization studies to show that its distribution is restricted to the core, a finding consistent with the premise that 110K is a component of the cross filaments (Matsudaira and Burgess, 1979). However, Coudrier et al. (1981) report that antibodies to 110K also stain the terminal web region even though cross filaments are lacking from rootlets.

The calmodulin associated with 110K is a major core protein. The molar ratio of calmodulin to actin makes it, after actin, the most abundant microvillar core protein (Howe et al., 1980; Glenney and Weber, 1980). Early on, calmodulin was localized in epithelial cell cytoskeletons to the microvillar cores (Glenney and Weber, 1980). Most of the calmodulin appears to be associated with 110K in a Ca^{++}-dependent manner (Glenney et al., 1980; Glenney and Weber, 1980; Glenney and Glenney, 1985) as was shown by ^{125}I-calmodulin gel overlays. More recent data indicate that the molar ratio of calmodulin to 110K is at least 4:1 (Swanljung-Collins and Collins, 1985), suggesting that some calmodulin is lost during core isolation.

Four groups have had recent success with purification and characterization of the 110K-calmodulin complex (Howe and Mooseker, 1983; Glenney and Glenney, 1984a; Collins and Borysenko, 1984; Verner and Bretscher, 1985). Solubilization of the complex apparently requires divalent cations and nucleoside triphosphate. A mild controversy has arisen regarding whether 110K is an integral membrane protein due to Glenney and Glenney's (1984) report that (1) 110K partitions into the detergent phase of Triton X-114; (2) it incorporates into liposomes; (3) it is solubilized by detergent; and (4) it is labeled by a hydrophobic membrane probe. All of these properties, with the exception of the last, were performed on 110K exposed to sodium dodecyl sulfate (SDS), perhaps leading to exposure of hydrophobic domains. On the other hand, Verner and Bretscher (1985) have shown that 110K-calmodulin will remain water soluble in the presence of excess adenosine triphosphate (ATP) during purification. Once pure the complex is stable even in the absence of ATP. Taken together these studies indicate that 110K-calmodulin has a hydrophobic domain, possibly membrane associated, but is a cytoplasmic protein. Consistent with the argument that 110K is not an integral membrane protein is the report of Cowell and Danielson (1984), who showed that 110K is synthesized on free polysomes.

The morphological and dissociation studies of Matsudaira and Burgess (1979, 1982b) led to the prediction that 110K-calmodulin should bind to the membrane and to actin. A potential membrane-binding glycoprotein for 110K of 140-kd has been identified using the gel overlay technique (Coudrier et al., 1983), although this protein has not been purified. The isolated 110K-calmodulin complex has been shown to bind actin in the absence, but not in the presence, of ATP (Howe and Mooseker, 1983; Collins and Borysenko, 1984; Verner and Bretscher, 1985). However, no periodicity to the actin binding was noted, although based on the earlier structural studies (Matsudaira and Burgess, 1982b) it seems likely that the 33-nm periodicity depends upon the presence of a filament bundle of a precise size and perhaps an intact membrane possessing the 110K-binding protein.

Some of the most exciting current work on 110K-calmodulin has come out of Collins's lab (Collins and Borysenko, 1984; Swanljung-Collins and Collins, 1985). Using an improved purification procedure, they have found that the 110K-calmodulin complex has properties resembling two classes of ATPases. They showed that the complex (1) has a Ca^{++}-stimulated Mg^{++}-ATPase activity that is further stimulated by F-actin and (2) has a K^{+}, EDTA-ATPase activity (EDTA = ethylenediaminotetraacetate). Thus, enzymatically, it resembles both myosin and membrane Ca^{++}-transport ATPases. However, 110K does not demonstrate immunological cross-reactivity to myosin, nor does it have a similar peptide map. Interestingly, 110K can be phosphorylated, although as yet the roles of phosphorylation or of calmodulin in the complex are unknown.

d. Ezrin. One newly identified minor protein of the core is an 80-kd polypeptide named ezrin, which is found in ~0.7% the molar abundance of actin (Bretscher, 1983b). This protein was found to be monomeric and slightly elongated and does not appear to strongly bind to actin, fimbrin, or villin although a detailed analysis has yet to be done. Recent studies have shown that ezrin is homologous to a well-characterized tyrosine kinase substrate found in A431 cells whose activity is stimulated by epidermal growth factor (EGF) (Gould et al., 1986). Ezrin, or p81, is found by immunofluorescence to be widely distributed in a variety of cell types and is localized to microspikes, microvilli, and ruffles of cultured cells (Bretscher, 1983b; Gould et al., 1986). No function has been assigned to exrin.

C. The Core: Problems and Perspectives

There are several interesting unresolved questions about the microvillus core. Starting from the top, we still do not know the makeup of the electron-dense tip, which in thin section appears to have a paracrystalline structure. It is entirely possible that villin, with its capping and nucleating abilities, is the tip protein. The fact that the 110K-calmodulin complex, making up the cross-filament, has an ATPase activity raises the possibility of some form of microvillar movement mediated by 110K. Clearly more information is needed about this complex, including its membrane-binding proteins. The vast majority of the actin-binding data for all core proteins concerns the characteristics of individual proteins. However, the core is a complex of several actin-binding proteins acting in concert. We know little about the synergistic or competing effects of villin, fimbrin, 110K-calmodulin, and ezrin. Since phosphorylation is well known to affect the properties of various proteins, the fact that both ezrin and 110K can be phosphorylated suggests some potential modulation of effects. Further, there are probably several other key core proteins present in minor amounts yet to be characterized.

Villin seems to be uniquely localized to the brush border (either intestinal or kidney), whereas fimbrin and ezrin are readily found elsewhere, suggesting that villin might play some specific role in the brush border. It is well known that micromolar levels of free Ca^{++} induce solation of the core and vesiculation of the microvillar membrane likely due to villin (Glenney et al., 1980; Howe et al., 1980; Burgess and Prum, 1982; Verner and Bretscher, 1983). This response to Ca^{++} might play a physiological role to protect the epithelium— and the organism—from the enterotoxic bacteria that produce modified pili to mediate adherence to the mucosal membrane and cause diarrhea. One early response to such bacteria is the solation of microvilli (Boedeker, 1984), likely due to a breach in the membrane to allow Ca^{++} influx that would trigger villin's severing activity. If these bacteria were to remain bound to the solated membrane, they would pass through the digestive tract and thereby allow the renewing absorptive epithelium to recover.

III. THE TERMINAL WEB

A. Structure

The first consideration of the terminal web as a distinct ultrastructural component of the absorptive cell was by Palay and Karlin (1959), who considered it an overly thickened ectoplasmic gel layer found in all cells. This region, now known to be composed of an anastomosing meshwork of specific structural proteins, is in fact precisely organized into specific domains.

Hull and Staehelin (1979) first clearly demonstrated that the terminal web can be subdivided into three zones, defined by the kinds and densities of filaments. These three zones or layers, termed the apical, adherens, and basal zones, are at the planes of the three junctions of the apical junctional complex: the tight, intermediate, and desmosomal junctions, respectively.

The most obvious terminal web features are the microvillar core rootlets. In any one cell the rootlets all penetrate to the same depth of about 0.7 μm and terminate at the apical portion of the basal zone (Burgess and Prum, 1982). Early ultrastructural studies suggested that the rootlet filaments splay apart (McNabb and Sandborn, 1964; Brunser and Luft, 1970; Mukherjee and Staehelin, 1971; Hull and Staehelin, 1979). However, more recent thin and thick serial sectioning (Matsudaira and Burgess, 1982c) and elegant quick-freeze deep-etch images (Hirokawa et al., 1982, 1983a,b; Hirokawa and Heuser, 1981) argue convincingly that splaying of rootlet filaments does not occur and that all filaments within a rootlet end at the same plane. These studies also show that the actin filaments are continuous from the tip of the microvillus to the base of the rootlet. That rootlet splaying is unlikely has significant implications for various models of brush-border motility to be discussed later.

Immunolocalization studies suggest that in addition to actin, the rootlets possess villin, α-actinin, myosin, tropomyosin, and caldesmon (Drenckhahn et al., 1983; Drenckhahn and Groschell-Stewart, 1980; Herman and Pollard, 1981; Geiger et al., 1979, 1981; Bretscher and Lynch, 1985). Interestingly, α-actinin, myosin, tropomyosin, and caldesmon, while associated with the rootlet portion of the actin bundle, are absent from that portion of the bundle within the core. One of the significant unresolved issues in the field is the restricted domain occupied by such proteins as tropomyosin and caldesmon. However, since rootlets have not been isolated, as have microvilli, no biochemical confirmation of the immunolocalization studies is present. Therefore, the assignment of protein locations to the rootlets and the terminal web in general by immunolocalization is subject to the caveat that no major redistribution, loss, or adventitious binding occurs during specimen preparation.

The population of filament proteins appears to be distinct in the inter-rootlet region of each zone with ~5-nm (in diameter) fine filaments populating the apical zone, ~8-nm fibrils predominant in the adherens zone, and a dense meshwork of 10-nm intermediate filaments present in the basal zone. A careful examination of the makeup of the inter-rootlet filaments reveals that they are not made up of actin (Begg et al., 1978; Hirokawa and Heuser, 1981; Hirokawa et al., 1982, 1983a). The apical zone 5-nm fine filaments are most likely composed of a spectrin-like protein termed TW 260/240 (Hirokawa et al., 1983a), whereas the 8-nm fibrils of the adherens zone are probably composed of myosin monomers or dimers (Hirokawa et al., 1982). It is likely, however, that both proteins are present in each zone, although each may have a preferential distribution (Glenney and Glenney, 1983a,b). The 10-nm filaments of the basal zone are intermediate filaments of the keratin family associated with the desmosomes (Franke et al., 1979); these filaments may have some association with TW 260/240 or the bases of the rootlets (Hirokawa et al., 1983).

The other prominent feature of the terminal web is the circumferential ring of filaments associated with the adherens junction. These filaments were first made note of by Farquhar and Palade (1963) and were clearly shown to be circumferentially organized thin filaments running parallel to the adherens junction by Hull and Staehelin (1979). The adherens zone ring is composed of actin filaments of mixed polarity (Burgess and Prum, 1982; Hirokawa et al., 1983b) associated with myosin, α-actinin, tropomyosin, and vinculin (Bretscher and Weber, 1978a; Mooseker et al., 1978; Craig and Pardo, 1979; Geiger et al., 1981; Herman and Pollard, 1981; Burgess and Prum, 1982). This ring mediates *in vitro* brush-border motility and will be discussed later.

Several other proteins associated with the cytoskeleton have been localized or functionally assigned to the terminal web without being precisely localized to a specific domain. These include myosin light-chain kinase (MLCK) and myosin light-chain phosphatase (MLCP) (Broschat et al., 1983; Keller et al., 1985), calpactin (Glenney and Tack, 1985), a tyrosine kinase (Gerke and Weber, 1984), vitamin D-dependent intestinal calcium-binding protein (CaBP) (Nemere and Norman, 1982), filamin (Bretscher and Weber, 1978b), and gelsolin (Yin et al., 1981).

Some of the terminal web proteins have been the subject of intensive investigation and will be discussed in some detail, whereas other proteins have only been identified as brush-border proteins by immunolocalization studies and will be considered briefly in functional terms. Included in the latter group are filamin, gelsolin, α-actinin, and vinculin. Another group of proteins, such as MLCK, MLCP, and calpactin kinase are assumed to be present because of changes in their substrate and will be viewed by their effects.

B. Proteins

1. TW 260/240 (Spectrin)

Spectrin is a tetrameric fibrous protein found associated with actin, ankyrin, and band 4.1 making up the latticework of the red blood cell membrane. In the last few years non-red blood cell spectrins have been found in several cell types including brain, where it is termed fodrin, and the epithelial brush border, where it is named TW 260/240. Glenney et al. (1982a) first identified and localized TW 260/240 in the terminal web. The protein has subsequently been shown to be a tetramer of two dissimilar high-molecular-weight subunits (260 and 240 kd) forming double-stranded molecules ~260 nm in length. Features of TW 260/240 include calcium-dependent binding to calmodulin (Glenney and Weber, 1980; Glenney et al., 1982b) and actin binding by the complex (Glenney et al., 1983; Pearl et al., 1984). However, *in vitro* studies to date raise serious questions about the physiological properties of TW 260/240 because it only binds actin under unphysiologically low salt conditions and, also unique to the spectrin family, it does not bind to either inside-out red blood cell vesicles or to ankyrin (Howe et al., 1985). Incidentally, no one has reported the presence of either ankrin or band 4.1 in the brush border. Interestingly, peptide mapping and antibody cross-reactivity studies demonstrate that the 240-kd subunits of avian spectrin, fodrin, and TW 260/240 are highly related to one another but distinct from their mammalian 240-kd counterparts and further show that the 240-kd mammalian subunits are all distinct (Glenney and Glenney, 1983b, 1984b). Therefore, current *in vitro* work does not support the widely ascribed properties of TW 260/240 being a rootlet crosslinking or a rootlet-membrane-linking complex. Unfortunately, brush-border tropomyosin, associated with rootlet actin, does not enhance TW 260/240's actin-binding properties (Burgess et al., 1987). It appears likely that some other rootlet or membrane protein will affect TW 260/240's binding properties and bring some physiological relevance to this important protein.

2. The Circumferential Ring, Myosin, and Motility

The concept of brush-border motility is an old one fueled by claims of *in vivo* movements (Thuneberg and Rostgaard, 1969; Sandstrom, 1971). While these claims remain unsubstantiated in spite of long hours of viewing with both conventional and video-enhanced light microscopy, it is now clear that brush borders readily move *in vitro*. Two key papers appeared in 1976 in which ATP-dependent motility of isolated brush borders was reported (Rodewald et al., 1976; Mooseker, 1976). The interpretations made by these authors were guided by the information available at that time (which included the impression, later shown not to be the case, that the rootlets splay apart) and hampered by a lack of knowledge of villin. Using demembranated brush borders Mooseker (1976) reported Ca^{++}-ATP dependent retraction of microvillar cores into the terminal web and proposed that rootlet actin splayed apart to interact with terminal web myosin. On the other hand, Rodewald et al. (1976) used membranated brush borders to detect Ca^{++}-independent radial contraction of the terminal web, leading to the hypothesis of rootlet actin splaying to interact with myosin and, importantly, that actin associated with the zonula adherens takes part in the contraction with rootlet actin. The subsequent work on villin, the careful description of the terminal web by Hull and Staehelin (1979), and the report of ATP-dependent *in vitro* contraction of an adherens circumferential ring in cultured retinal cells (Owirabe et al., 1981) led to a reinvestigation of brush border motility (Burgess and Prum, 1982). We found that the Ca^{++} effect was due to solation of the microvillus cores caused by villin, leaving the rootlets intact. This work demonstrated that microvilli are *not* retracted into the terminal web. We also made note of the adherens circumferential ring, of it being composed of actin and myosin, and suggested that brush-border motility is caused by contraction of the adherens ring.

Evidence in support of terminal web contraction mediated by the circumferential ring came in reactivation studies using intact sheets of glycerinated epithelium (Burgess, 1982). I found that ATP-dependent contractions occurred with enough force to pull apart the junctional complex. At the same time another report appeared in which terminal web contractions were noted in isolated brush borders and were attributed to inter-rootlet contractions with the myosin in the terminal web (Keller and Mooseker, 1982). Interestingly, neither study found that contraction required Ca^{++}, but only that ATP was required. We argue against inter-rootlet contractions because (1) inter-rootlet distances do not appear to decrease and (2) slight contractions cause only the microvilli adjacent to the ring to fan out. A basic consensus that terminal web contractions are mediated by the adherens ring came about when Hirokawa et al. (1983b) independently reproduced our earlier work on reactivations of glycerinated sheets of epithelium.

The regulation of brush border motility by myosin phosphorylation has been extensively studied. Keller and Mooseker (1982) detected phosphorylation of a 20-kd polypeptide, whose phosphorylation was enhanced by Ca^{++}, which was likely myosin light chain. The use of one-dimensional gels in this study made it difficult to determine with certainty whether this polypeptide was in fact the 20-kd myosin light chain because brush-border preparations are contaminated with histones which can be phosphorylated. Direct demonstration of the dependence of brush contraction on myosin light-chain phosphorylation came from our studies in which (1) ATPγS, a substrate for MLCK but not for the ATPase, could trigger phosphorylation but not contraction; (2) inosine triphosphate (ITP), an ATPase substrate but not a substrate for the kinase, did not trigger contraction; and (3) the combination of ATPγS and ITP triggered contraction (Broschat et al., 1983). We also suggested, consistent with the quick-freeze, deep-etch studies, that myosin is unassembled in the terminal web until phosphorylated, but undergoes assembly for contraction triggered by MLCK. Similar results have appeared recently, although there is some debate as to the association of myosin to the cytoskeleton (Keller et al., 1985).

We and others have recently examined, in some detail, brush border myosin. We found that the 20-kd myosin light chain is phosphorylated on both serine and threonine in isolated brush borders and upon phosphorylation its ATPase activity is enhanced (Broschat and Burgess, 1984, 1986a). The ATPase activity is also actin activated and it is further enhanced when the actin is saturated with brush border tropomyosin. Researchers in Collins's laboratory have identified two Ca^{++}-calmodulin-dependent MLCKs, one Ca^{++}-calmodulin-dependent heavy-chain kinase, and one Ca^{++}-independent MLCK (Rieker et al., 1985). The Ca^{++}-dependent kinase phosphorylates the light chain on serine, whereas the Ca^{++}-dependent kinase phosphorylates a threonine. These workers also note enhancement of ATPase activity upon phosphorylation. The foregoing studies seem to explain the lack of a Ca^{++} requirement or brush border motility, although no role for heavy-chain phosphorylation is known at this time.

The role of brush-border motility *in vivo* is not known. Most investigators of brush-border motility *in vitro* speculate along the same lines as to potential functions. Central among these ideas is the thought that adherens ring contractions facilitate nutrient transport by altering the strandedness of the tight junction and thereby the transepithelial resistance. Such perturbants as changing the osmotic load (Madara, 1983), increased intracellular cAMP (adenosine 3′:5′-cyclic monophosphate) (Duffer et al., 1981), or increased mechanical tension (Pitelka and Taggart, 1983) alter the orientation of tight junction strands and/or the transepithelial resistance, findings consistent with a role for a contracting adherens ring. Moreover, earlier work on morphogenesis of previllus ridges suggests that the adherens ring acts to constrict the apical ends of cells to fold the epithelium as early as 8–9 days of incubation, which is well before villi appear on day 16 (Burgess, 1975). Such a result indicates that the ring may play roles in maintaining the epithelium in a paddle-shaped form in the adult or in sealing the epithelium as dying cells slough off the tips of mature villi.

3. Tropomyosin

Tropomyosin is known to be present in both the circumferential ring and the rootlets. We have recently characterized intestinal epithelial cell tropomyosin (Broschat and Burgess, 1986b), extending a preliminary report on its purification (Mooseker et al., 1982a). This tropomyosin was found to be composed of three distinct isoforms of about 32 kd. It cooperatively binds to F-actin under physiological salt and Mg^{++} concentrations with an apparent K_a of $8 \times 10^6\ M^{-1}$, which is much higher than that of other nonmuscle tropomyosins. It was also found to bind stoichiometrically to F-actin at physiological ionic strength even in the absence of Mg^{++}. All of its functional properties appear more related to smooth muscle than to other low molecular weight nonmuscle tropomyosins.

Functionally, tropomyosin is known to protect actin filaments from the Ca^{++}-activated F-actin-severing abilities of villin, a property likely explaining the stability of the rootlets under levels of free Ca^{++} known to disrupt microvillar cores (Burgess and Prum, 1982). By making the rootlets stable, the cell has a template for microvillar regrowth in the event of microvillar disruption that can be caused by various enterotoxic bacteria. The presence of tropomyosin in the terminal web circumferential ring suggests a second function for the protein. We have found that brush-border tropomyosin stimulates the actin-activated ATPase of phosphorylated brush-border myosin (Broschat et al., 1985). This result indicates that tropomyosin plays a regulatory role for brush border motility.

The major unresolved question regarding the distribution of tropomyosin in the brush border is why it is absent from the core filaments. Our findings, discussed earlier, that fully saturated villin bundles can still bind saturating amounts of tropomyosin does not resolve this issue. Perhaps it is the combination of villin with fimbrin that prevents tropomyosin from binding to the core. Another possibility suggested by immunofluorescence data (Bretscher and Lynch, 1985) is that caldesmon might enhance tropomyosin's binding to the rootlets. The fact that much tropomyosin remains associated with the brush border, while large amounts are solubilized during cell disruption in the absence of Mg^{++} and salt, conditions unfavorable for actin-tropomyosin binding, further supports the idea of protein-protein interactions strengthening the association of tropomyosin to the terminal web. Only further work will resolve the issue of how tropomyosin and other proteins are regionalized in the brush border.

4. Calpactin

Calpactin is a recently described terminal web Ca^{++}-regulated actin- and spectrin-binding protein that serves as a major substrate for tyrosine kinases and is a member of a large family of such proteins. As first analyzed by Gerke and Weber (1984, 1985) intestinal calpactin is a complex composed of two copies of a 10-kd subunit and two copies of a 36-kd subunit. The 10-kd subunit is homologous to Ca^{++}-binding brain S-100 protein (Glenney and Tack, 1985). Initial actin- and spectrin-binding and phosphorylation studies revealed that unphysiologically high Ca^{++} levels were required for such effects (Gerke and Weber, 1984; Glenney and Glenney, 1985; Glenney, 1985). More recent experiments have revealed that phospholipid, especially phosphoserine, brings both 36-kd Ca^{++}-binding activity and Ca^{++}-stimulated 36-kd tyrosine phosphorylation by $pp60^{src}$ to micromolar free Ca^{++} levels (Glenney, 1985, 1986a,b). The intestinal calpactin appears to be distinct from—but related to—other tyrosine kinase substrates of the same size (Glenney, 1986c).

While both 36-kd Ca^{++}-binding and phosphorylation are brought into physiologically significant ranges by phospholipid, work to date has not commented on actin and spectrin binding. It seems likely that the Ca^{++} requirements for actin and TW 260/240 will fall into the physiological range in the presence of phospholipid and, if it does, then calpactin will be a leading candidate as the protein linking adherens ring and rootlet actin and TW 260/240 to the membrane. However, the role of the tyrosine kinase in the properties of intestinal calpactin is yet to be resolved.

C. The Terminal Web: Problems and Perspectives

Major dissection of the terminal web is yet to be accomplished. The fact that the rootlets, circumferential ring, and the inter-rootlet regions have yet to be isolated means that the factors mediating such interactions as those maintaining TW 260/240, caldesmon, and tropomyosin binding remain unknown. We do not understand the interactions which control the associations of myosin, calpactin, MLCK, MLCP, and caldesmon to the cytoskeleton. Clearly the major unresolved questions concern the regulation of the restricted domains occupied by the many cytoskeletal proteins. Only when thorough recombination studies are done using several proteins at once will a better understanding of terminal web structure come about. The circumferential adherens ring structure raises many questions. First is how and in what form myosin is associated with the ring. The physiological relevance of the ring remains enigmatic, as does all regulatory aspects of its contraction. We do not understand how the ring remains bundled or how it is associated with the membrane. Major questions still remain about the suggestions that there are multiple myosin isoforms, one for use in the ring and one for vesicle movement (Keller et al., 1985). No direct evidence is at hand about such isoforms. Further, the role of inter-rootlet myosin is not yet known. Finally, we do not know if there is any interaction between the terminal web rootlets, or TW 260/240, and the underlying keratin thick filament system.

IV. BRUSH BORDER DEVELOPMENT AND MICROVILLUS GROWTH

A. Epithelial Development

The arrangement of the brush border cytoskeleton into specific domains and the constancy in length of microvilli raises considerable organizational questions from a developmental viewpoint. Since the adult epithelium is constantly renewed through division of undifferentiated crypt cells, development proceeds as cells migrate out of the crypts. When cells reach the tops of crypts, they demonstrate both specific ultrastructural features and specific integral membrane glycoproteins such as alkaline phosphatase and sucrose-isomaltase (Trier and Madara, 1981). Most of these studies have focused on specific enzymes or antigens as they appear on cells when they reach the base of the villus (Quaroni, 1985a). Very little is known about the factors that restrict cell division to the crypts or regulate differentiation of the various cell types in the mature epithelium, or about such controlling factors as cell or tissue interactions.

However, much is known about differentiation regulation in the embryonic and neonatal epithelium prior to the time sloughing at the villus tip begins. Early morphological work showed that there are wide differences between villus formation in mammals vs. chickens or other vertebrates (Hilton, 1902). Since most of the work on the brush border cytoskeleton has been done with chickens, I will describe development of the chicken intestine—although there has been equally detailed work done on mice and rat intestines.

Villus formation in chickens occurs by folding of the simple epithelium first into 16 longitudinal previllus ridges, which are then folded into a zigzag pattern; out of the corner of each zig and zag a villus elongates during the last quarter of incubation (Coulombre and Coulombre, 1958; Grey, 1972). As the villi begin to grow, on day 16 of a 21-day incubation period, the columnar cells begin to flatten their apical surfaces and then increase in height (Hinni and Watterson, 1963; Grey, 1972). As the villi grow out mitoses become gradually restricted to the future crypt region (Overton and Shoup, 1964; Bellware and Betz, 1970), and by two days after hatching cell division is restricted to the crypts and sloughing begins (Hinni and Watterson, 1963). The microvilli also undergo morphological differentiation during this period by achieving their adult density by day 20 of incubation and their adult length 3 days after hatching (Overton and Shoup, 1964; Chambers and Grey, 1979; Stidwill and Burgess, 1986). Accompanying this morphological maturation is the coincident appearance of specific luminal membrane enzymes (Moog, 1950, 1979).

Several lines of evidence suggest that chicken villus and brush-border development is under hormonal regulation (Moog, 1979). First, the hormone hydrocortisone has been shown to cause precocious brush border differentiation (Moog and Richardson, 1955). Second, inhibition of thyroxine release retards villus growth (Moog, 1961). Third, hypophysectomy arrests duodenal differentiation at an immature form (Hinni and Watterson, 1963; Bellware and Betz, 1970). However, cultured explants of embryonic intestine demonstrate hydrocortisone-independent acceleration of differentiation that is further stimulated by thyroxine (Black, 1978; Black and Moog, 1978). Similar kinds of studies also show hormonal control of differentiation of mammalian intestines.

B. The Cytoskeleton in Brush Border Development

There has been a renewed interest in the role of the cytoskeleton in morphogenesis of both the epithelium and the brush border. The terminal web adherens circumferential ring has been shown to be present during early development of the epithelium and has been suggested to mediate cell-shape changes responsible for folding the epithelium into previllus ridges (Burgess, 1975). The ring was found to be disrupted by cytochalasin B, which also blocked folding of the epithelium. This study clearly indicates that the terminal web circumferential ring—and therefore ring actin, myosin, and associated proteins—functions to constrict the apical ends of the absorptive cells early in development, before the time the brush border is present.

Structural features of microvillus growth and core-filament assembly during development have been investigated using several different species. The consensus of studies (Overton and Shoup, 1964; Overton, 1965; Overton et al., 1965; Chambers and Grey, 1979; Stidwill and Burgess, 1986) finds the following sequence of developmental events for brush-border growth: (1) From 5 to 10 days of incubation microvilli are sparse and the surfaces of the cells are ballooned into the lumen of the gut. (2) The ballooned apical cell surfaces are retracted as the cell surface becomes dome shaped on day 11 and remain so until the terminal web begins forming on day 19. Numerous microvilli, averaging 0.5 μm in length, are present and arranged at various angles to the cell surface at this time. (3) Rootlets begin to elongate on day 10 and are fully formed by day 15. (4) The adult number of microvilli is present on day 21 (hatching), and the mature number of core filaments occurs 2 days later. (5) The terminal web stratifies between day 19 and 3 days after hatching. (6) Finally, microvilli undergo a major burst in elongation between days 2 and 4 after hatching to achieve the adult length.

Several studies provide some hints as to how core filaments become arranged during development. Based on the careful observations of Chambers and Grey (1979) it appears that intestinal microvillar core asembly proceeds with very loosely organized filaments first appearing parallel and adjacent to the membrane, leaving the internal region of the core vacant of filaments. Early reports suggested that core filaments become associated with the membrane only where dense plaques were present (Bonneville and Weinstock, 1970; Tilney and Cardell, 1970), although Chambers and Grey (1979) found that core filaments attached to the membrane in regions lacking a plaque. Core filaments then appear to fill in the center of the core, and we have observed that before day 20 the filament bundle is more loosely arranged than the near-hexagonal packing that occurs by day 2.

The organization and role of the cytoskeleton during microvillus growth and in maintenance of normal microvillus length in the developing intestine has been studied extensively (Overton and Shoup, 1964; Chambers and Grey, 1979; Stidwill and Burgess, 1986). In the adult and the developing chick embryo, intestinal microvilli are of a characteristic length (Burgess and Prum, 1982; Stidwill and Burgess, 1986). One major question, still unresolved, is how this specific, finite length is regulated. This question is made more significant because all core cytoskeletal proteins are turning over, as has been shown for adult epithelial cells from both chickens and pigs (Stidwill et al., 1984; Cowell and Danielsen, 1984). Our studies documented relative rates of turnover of brush border cytoskeletal proteins, and we found that the microvillar core proteins actin, villin, and fimbrin turn over as a cohort at a slower rate than do 110K and TW 260/240. More recently, Cowell and Danielsen (1984) have shown very rapid incorporation of label into villin and 110K, suggesting a high turnover rate, more rapid than that of membrane proteins. The significant result from these studies is that the cytoskeleton is not a static structure but, rather, that there must be a constant addition of new proteins and a degradation of old proteins.

Further evidence of the dynamic state of the cytoskeleton is educed from the observations that inhibition of protein synthesis (LeCount and Grey, 1972), fasting of animals (Misch et al., 1980), and colchicine treatment (Buschmann, 1983) all cause shortening of microvilli. These studies, coupled with the experiments demonstrating turnover of cytoskeletal proteins, appear to indicate that the constancy in length of microvilli is due to core protein turnover in a steady-state condition.

More recently, we have found evidence that microvillus length is likely regulated by the interplay of steady-state turnover with the G- to F-actin ratio (Stidwill and Burgess, 1986). We and others found that microvilli undergo a burst in elongation 2–4 days after hatching. By determining the number of microvilli, the number of core filaments, the microvillus lengths, and the cell G- to F-actin ratio during development we determined that an increase in G-actin probably drives assembly of more F-actin and thus causes longer microvilli. We found that microvilli from embryos and adults are both constant in length, with those from the embryo being much shorter, but that the cell G- to F-actin ratio is the same in both cases. Just before microvilli begin to grow, the G-actin content dramatically rises, consistent with the hypothesis that a fixed ratio of G- to F-actin maintains mirovillus height. Consistent with this model are the results of LeCount and Grey (1972), who found that inhibition of protein synthesis, which would likely deplete G-actin levels, caused a shortening of microvilli. Moreover, Tilney and Cardell (1970) showed that high hydrostatic pressure caused the resorption of microvilli and the disruption of their cores; then, upon release of pressure, the cores reform and the microvilli regrow. The pressure likely drove the actin equilibrium to the G-form, and upon pressure release the high G-actin levels drove rapid filament reassembly.

We have initiated a series of studies on the regulation of expression of specific brush-border cytoskeletal genes during development. Using cDNA probes for β-actin, spectrin, and tropomyosin we found high levels of both tropomyosin and spectrin messages on days 18 and 19 of incubation, whereas the peak level of actin message occurred 2 days after hatching (Obenauf and Burgess, 1986). The period of highest levels of actin message correlate well with the time wherein increased levels of G-actin were detected biochemically (Stidwill and Burgess, 1986) and suggest that the increased levels of G-actin that are likely to drive microvillus elongation are under transcriptional control. The highest levels of spectrin and tropomyosin expression correlate well with the time wherein the terminal web is assembling. It will be interesting to determine if the genes for other microvillar core proteins are coordinately expressed with actin during microvillus elongation.

C. Culture Systems for Brush Border Development

Considerable effort has gone into the development of organ and cell culture systems that ideally would provide dividing and differentiating crypt cells. Organ cultures of adult fragments have been difficult to maintain due to rapid cell death. On the other hand, cultures of embryonic or fetal intestinal fragments have been more successful. In this regard, the validity of studying the differentiation of fetal or embryonic intestinal epithelium is strengthened by recent experiments showing that fetal epithelial cells have similar luminal membrane antigens to adult crypt cells (Quaroni, 1985a, 1986).

Successful cultures of fetal or embryonic rat (Quaroni, 1985b; DeRitis et al., 1975) and chick (Kato, 1970; Black and Moog, 1978; Corradino, 1979) organ fragments have been reported which, as discussed above, demonstrate hormone-dependent differentiation. The fetal intestinal organ culture system of Quaroni (1985b), which consists of vesicles of right-side-out epithelium, appears to show great promise for future studies. On the other hand, monolayer cultures of isolated crypts have not been as promising.

Recently, Zweibaum and colleagues, among others, have taken advantage of the human colon carcinoma cell lines HT-29 and Caco-2 that undergo differentiation in culture (Zweibaum et al., 1983, 1985; Pinto et al., 1982, 1983). They have selected a subpopulation of HT-29 cells that grow in sugar-free medium and that differentiate, as monitored by brush border formation and sucrose-isomaltase appearance, in the absence of glucose. Because differentiation can be halted and reversed by glucose addition, this system holds much promise for the experimental analysis of brush border differentiation.

D. Development: Problems and Perspectives

Probably less is known about microvillus and brush border development than about any other aspect of the cytoskeleton. The list of questions is long, leaving much room for new investigations. Nothing is known about the way filaments are originally associated with the membrane; how the exact number of core filaments or microvilli, which are precisely controlled in development, are determined; or how bundling of filaments changes during development. The controlling signals for terminal web maturation, core filament growth, and microvillus length and number regulation are not known. One interesting and lingering question concerns which end of the filament bundle monomers add onto. Mooseker et al. (1982b) found that both ends of the microvillus core were competent for assembly in isolated membranated brush borders. The only experiments using intact cells that might have some bearing on assembly are the pressure experiments of Tilney and Cardell (1970) and our early studies on the effects of cytochalasin B on embryonic epithelia (Burgess and Grey, 1974; see also Noda, 1985). We found that cytochalasin B, now known to block the fast end for assembly of actin filaments, which is also the end attached to the dense plaque, caused rootlet growth. One interpretation of these results is that assembly was promoted off the slow end for assembly, probably because the concentration of monomers was high in these embryonic cells. Therefore, the *in vitro* and *in vivo* observations are consistent with monomer addition onto either end of the core. However, no definitive evidence is at hand to solve this interesting question. Significantly, we have little knowledge about the temporal or spatial appearance of the many cytoskeletal proteins during brush border development or how their expression is coordinated. It is likely that only through more molecular methodologies will these questions be approached.

While the various *in vitro* systems, including the HT-29 cell line, show great potential, much basic groundwork must be done. Such standard work as investigation of the ultrastructure of both microvillus and brush-border growth must be carried out to determine whether development *in vitro* mimics development *in vivo*. The fact that rootlet growth, terminal web growth, and microvillus elongation are uncoupled *in vivo* may be very pertinent when development *in vitro* is compared using these cell lines. However, their great potential as an experimental system should lead to significant new insight on the molecular control of brush border formation.

ACKNOWLEDGMENTS

I wish to thank Drs. Tony Bretscher, Jimmy Collins, John Glenney, Paul Matsudaira, and Mark Mooseker for sharing unpublished work, manuscripts, and preprints and for many useful discussions. I also thank Drs. Kay Broschat and Steven Obenauf for reading the manuscript critically and for many good suggestions. Great appreciation is given to Mrs. Maria Martinez for her skill in preparing this manuscript. Finally, I thank Ms. Pat Veno for the excellent illustration. The original work from my lab reported here was supported by National Institutes of Health Grant DK 31643.

REFERENCES

Begg, D. A., Rodewald, R., and Rebhun, L. I. (1978) J. Cell Biol. *79:* 846–852.

Bellware, F. T., and Betz, T. W. (1970) J. Embryol. Exp. Morphol. *24:* 335–355.

Black, B. L. (1978) Am. J. Anat. *153:* 573–600.

Black, B. L., and Moog, F. (1978) Dev. Biol. *66:* 232–249.

Boedeker, E. C., ed. (1984) *Attachment of Organisms to the Gut Mucosa.* CRC Press, Boca Raton, FL.

Bonder, E., and Mooseker, M. (1983) J. Cell Biol. *96:* 1097–1107.

Bonneville, M. A., and Weinstock, M. (1970) J. Cell Biol. *44:* 151–171.

Bretscher, A. (1981) Proc. Natl. Acad. Sci. USA *78:* 6849–6853.

Bretscher, A. (1983a) in *Cell and Muscle Motility* (J. Shay, ed.), Vol. 4, pp. 239–268. Plenum, New York.

Bretscher, A. (1983b) J. Cell Biol. *97:* 425–432.

Bretscher, A., and Lynch, A. (1985) J. Cell Biol. *100:* 1656–1663.

Bretscher, A., and Weber, K. (1978a) Exp. Cell Res. *116:* 397–407.

Bretscher, A., and Weber, K. (1978b) J. Cell Biol. *79:* 839–845.
Bretscher, A., and Weber, K. (1979) Proc. Natl. Acad. Sci. USA *76:* 2321–2325.
Bretscher, A., and Weber, K. (1980a) J. Cell Biol. *86:* 335–340.
Bretscher, A., and Weber, K. (1980b) Cell *20:* 839–847.
Broschat, K. O., and Burgess, D. R. (1984) J. Cell Biol. *99:* 5a.
Broschat, K. O., and Burgess, D. R. (1986a) Ms. in preparation.
Broschat, K. O., and Burgess, D. R. (1986b) J. Biol. Chem. *261:* 13350–13359.
Broschat, K. O., Stidwill, R. P., and Burgess, D. R. (1983) Cell *35:* 561–571.
Broschat, K. O., Shoreibah, M., and Burgess, D. R. (1985) J. Cell Biol. *101:* 403a.
Brown, A. L. (1962) J. Cell Biol. *12:* 623–627.
Brunser, O., and Luft, J. H. (1970) J. Ultrastruct. Res. *31:* 291–311.
Burgess, D. R. (1975) J. Embryol. Exptl. Morphol. *34:* 723–740.
Burgess, D. R. (1982) J. Cell Biol. *95:* 853–866.
Burgess, D. R., and Grey, R. D. (1974) J. Cell Biol. *62:* 566–574.
Burgess, D. R., and Prum, B. F. (1982) J. Cell Biol. *94:* 97–107.
Burgess, D. R., Broschat, K. O., and Hayden, J. (1987) J. Cell Biol. *104:* 29–40.
Buschmann, R. J. (1983) Cell Tissue Res. *231:* 298–299.
Chambers, C., and Grey, R. D. (1979) Cell Tissue Res. *204:* 387–405.
Collins, J. H., and Borysenko, C. W. (1984) J. Biol. Chem. *259:* 14128–14135.
Corradino, R. A. (1979) Arch. Biochem. Biophys. *192:* 302–310.
Coudrier, E., Reggio, H., and Louvard, D. (1981) J. Mol. Biol. *152:* 49–66.
Coudrier, E., Reggio, H., and Louvard, D. (1983) EMBO J. *2:* 469–475.
Coulombre, A. J., and Coulombre, J. L. (1958) J. Embryol. Exptl. Morphol. *6:* 403–411.
Cowell, G. M., and Danielsen, E. M. (1984) FEBS (Fed. Eur. Biochem. Soc.) Lett. *172:* 309–314.
Craig, S. W., and Lancashire, C. L. (1980) J. Cell Biol. *84:* 655–667.
Craig, S. W., and Pardo, J. V. (1979) J. Cell Biol. *80:* 203–210.
Craig, S. W., and Powell, L. D. (1980) Cell *22:* 739–746.
Cribbs, D. H., Glenney, J. R., Kaulfus, P., Weber, K., and Lin, S. (1982) J. Biol. Chem. *257:* 395–399.
Dalton, A. J. (1951) Am. J. Anat. *89:* 109–133.
Danielsen, E. M., Cowell, G. M., Noren, O., and Sjostrom, H. (1984) Biochem. J. *221:* 1–14.
DeRitis, G., Falchuk, Z. M., and Trier, J. S. (1975) Dev. Biol. *45:* 304–317.
Drenckhahn, D., and Groschel-Stewart, U. (1980) J. Cell Biol. *86:* 475–482.
Drenckhahn, D., Hofmann, H.-D., and Mannberg, H. (1983) Cell Tissue Res. *278:* 409–414.
Duffy, M. E., Hainau, B., Ho, S., and Bentzel, C. J. (1981) Nature (Lond.) *204:* 451–453.
Farquhar, M. G., and Palade, G. E. (1963) J. Cell Biol. *17:* 375–412.
Franke, W., Appelhans, B., Schmid, E., Freudenstein, C., Osborn, M., and Weber, K. (1979) Eur. J. Cell Biol. *19:* 255–268.
Geiger, B., Tokuyasu, K. T., and Singer, S. J. (1979) Proc. Natl. Acad. Sci. US *76:* 2833–2837.
Geiger, B., Dutton, A., Tokuyasu, K. T., and Singer, S. J. (1981) J. Cell Biol. *91:* 614–628.
Gerke, V., and Weber, K. (1984) EMBO J. *3:* 227–233.
Gerke, V., and Weber, K. (1985) J. Biol. Chem. *260:* 1668–1695.
Glenney, J. R. (1985) FEBS (Fed. Euro. Biochem. Soc.) Lett. *192:* 79–82.
Glenney, J. R. (1986a) FEBS (Fed. Eur. Biochem. Soc.) Lett. in press.
Glenney, J. R. (1986b) Exptl. Cell Res. *162:* 183–190.
Glenney, J. R. (1986c) Proc. Natl. Acad. Sci. USA *83:* 4258–4262.
Glenney, J. R., and Glenney, P. (1983a) Cell *34:* 503–512.
Glenney, J. R., and Glenney, P. (1983b) Cell Motil. *3:* 671–682.
Glenney, J. R., and Glenney, P. (1984a) Cell *37:* 743–751.
Glenney, J. R., and Glenney, P. (1984b) Eur. J. Biochem. *144:* 529–539.
Glenney, J. R., and Glenney, P. (1985) J. Cell Biol. *100:* 754–763.
Glenney, J. R., and Tack, B. F. (1985) Proc. Natl. Acad. Sci. USA *82:* 7884–7888.

Glenney, J. R., and Weber, K. (1980) J. Biol. Chem. *255:* 10551-10554.
Glenney, J. R., and Weber, K. (1981) Proc. Natl. Acad. Sci. USA *78:* 2810-2814.
Glenney, J. R., Bretscher, A., and Weber, K. (1980) Proc. Natl. Acad. Sci. USA *77:* 6458-6462.
Glenney, J. R., Geisler, N., Kaulfus, P., and Weber, K. (1981a) J. Biol. Chem. *256:* 8156-8161.
Glenney, J. R., Kaulfus, P., Matsudaira, P., and Weber, K. (1981b) J. Biol. Chem. *256:* 9283-9288.
Glenney, J. R., Kaulfus, P., and Weber, K. (1981c) Cell *24:* 471-480.
Glenney, J. R., Glenney, P., Osborn, M., and Weber, K. (1982a) Cell *28:* 843-854.
Glenney, J. R., Glenney, P., and Weber, K. (1982b) Proc. Natl. Acad. Sci. USA *79:* 4002-4005.
Glenney, J. R., Osborn, P., and Weber, K. (1982c) Exp. Cell Res. *138:* 199-205.
Glenney, J. R., Glenney, P., and Weber, K. (1983) J. Cell Biol. *96:* 1491-1496.
Gould, K. L., Cooper, J. A., Bretscher, A., and Hunter, T. (1986) J. Cell Biol. *102:* 660-669.
Granger, B., and Baker, R. F. (1950) Anat. Rec *107:* 423-441.
Grey, R. D. (1972) J. Morphol. *137:* 193-214.
Herman, I. M., and Pollard, T. D. (1981) J. Cell Biol. *86:* 346-351.
Hesterberg, L., and Weber, K. (1983a) J. Biol. Chem. *258:* 359-364.
Hesterberg, L., and Weber, K. (1983b) J. Biol. Chem. *258:* 365-369.
Hesterberg, L. K., and Weber, K. (1986) Eur. J. Biochem. *154:* 135-140.
Hilton, W. A. (1902) Am. J. Physiol. *1:* 459-504.
Hinni, J. B., and Watterson, R. L. (1963) J. Morphol. *113:* 381-427.
Hirokawa, N., and Heuser, J. (1981) J. Cell Biol. *91:* 399-409.
Hirokawa, N., Tilney, L. G., Fujiwara, K., and Heuser, J. E. (1982) J. Cell Biol. *94:* 425-443.
Hirokawa, N., Cheney, R. E., and Willard, M. (1983a) Cell *32:* 953-965.
Hirokawa, N., Keller, T. C. S., Chasen, R., and Mooseker, M. S. (1983b) J. Cell Biol. *96:* 1325-1336.
Howe, C. L., and Mooseker, M. S. (1983) J. Cell Biol. *97:* 974-985.
Howe, C. L., Mooseker, M. S., and Graves, T. A. (1980) J. Cell Biol. *85:* 916-923.
Howe, C. L., Sacramone, L. M., Mooseker, M. S., and Morrow, J. S. (1985) J. Cell Biol. *101:* 1379-1385.
Hull, B. E., and Staehelin, L. A. (1979) J. Cell Biol. *81:* 67-82.
Ishikawa, H., Bischoff, R., and Holtzer, H. (1969) J. Cell Biol. *43:* 312-328.
Ito, S. (1965) J. Cell Biol. *27:* 475-491.
Kato, Y. (1970) Transplantation *10:* 354-370.
Keller, T. C. S., and Mooseker, M. S. (1982) J. Cell Biol. *95:* 943-959.
Keller, T. C. S., Conzelman, K. A., Chasan, R., and Mooseker, M. S. (1985) J. Cell Biol. *100:* 1647-1655.
LeCount, T. S., and Grey, R. D. (1972) J. Cell Biol. *53:* 601-605.
McNabb, J. D., and Sandborn, E. (1964) J. Cell Biol. *22:* 701-704.
McNutt, N. S. (1978) J. Cell Biol. *79:* 774-787.
Madara, J. L. (1983) J. Cell Biol. *97:* 125-136.
Matsudaira, P. T., and Burgess, D. R. (1979) J. Cell Biol. *83:* 667-673.
Matsudaira, P. T., and Burgess, D. R. (1982a) J. Cell Biol. *92:* 648-656.
Matsudaira, P. T., and Burgess, D. R. (1982b) J. Cell Biol. *92:* 657-664.
Matsudaira, P. T., and Burgess, D. R. (1982c) Cold Spring Harbor Symp. Quant. Biol. *46:* 845-854.
Matsudaira, P. T., Mandelkow, E., Renner, W., Hesterberg, L. K., and Weber, K. (1983) Nature (Lond.) *301:* 209-221.
Matsudaira, P. T., Jakes, R., Camerson, L., and Atherton, E. (1985a) Proc. Natl. Acad. Sci. USA *82:* 6788-6792.
Matsudaira, P. T., Jakes, R., and Walker, J. E. (1985b) Nature (Lond.) *315:* 248-250.
Miller, D., and Crane, R. K. (1961) Biochem. Biophys. Acta *52:* 293-298.

Millington, P. F., and Finean, J. B. (1962) J. Cell Biol. *14:* 125–134.
Misch, P., Giebel, P., and Faust, R. (1980) Eur. J. Cell Biol. *21:* 264–279.
Moog, F. (1950) J. Exp. Zool. *115:* 109–130.
Moog, F. (1961) Gen. Comp. Endocrinol. *1:* 416–432.
Moog, F. (1979) in *Development of Mammalian Absorptive Processes*, Ciba Found. Symp. No. 70 (K. Elliot and J. Whelan, eds.), pp. 31–44, Exerpta Medica, Amsterdam.
Moog, F., and Richardson, D. (1955) J. Exp. Zool. *130:* 29–51.
Mooseker, M. S. (1976) J. Cell Biol. *71:* 417–432.
Mooseker, M. S. (1985) Annu. Rev. Cell Biol. *1:* 209–241.
Mooseker, M. S., and Stephens, R. E. (1980) J. Cell Biol. *86:* 466–474.
Mooseker, M. S., and Tilney, L. G. (1975) J. Cell Biol. *67:* 725–743.
Mooseker, M. S., Pollard, T. D., and Fujiwara, K. (1978) J. Cell Biol. *79:* 444–453.
Mooseker, M. S., Graves, T. A., Wharton, K. A., Falco, N., and Howe, C. L. (1980) J. Cell Biol. *87:* 809–822.
Mooseker, M. S., Bonder, E. M., Grimwade, B. G., Howe, C. L., and Keller, T. C. S. (1982a) Cold Spring Harbor Symp. Quant. Biol. *46:* 855–870.
Mooseker, M. S., Pollard, T. D., and Wharton, K. A. (1982b) J. Cell Biol. *95:* 222–233.
Mukherjee, T. M., and Staehelin, L. A. (1971) J. Cell Sci. *8:* 573–599.
Mukherjee, T. M., and Wynn Williams, A. (1967) J. Cell Biol. *34:* 447–461.
Nemere, I., and Norman, A. W. (1982) Biochim. Biophys. Acta *694:* 307–327.
Neutra, M. R. (1979) Anat. Rec. *193:* 367–381.
Noda, S. (1985) J. Embryol. Exptl. Morphol. *85:* 47–64.
Obenauf, S., and Burgess, D. R. (1986) Ms. in preparation.
Overton, J. (1965) J. Exp. Zool. *159:* 195–202.
Overton, J., and Shoup, J. (1964) J. Cell Biol. *21:* 75–85.
Overton, J., Eichholz, A., and Crane, R. K. (1965) J. Cell Biol. *26:* 693–706.
Owaribe, K., Kodama, R., and Eguchi, G. (1981) J. Cell Biol. *90:* 507–514.
Palay, S. L., and Karlin, L. J. (1959) J. Biophys. Biochem. Cytol. *5:* 363–372.
Pearl, M., Fishkind, D., Mooseker, M. S., Keene, D., and Keller, T. C. S. (1984) J. Cell Biol. *98:* 66–78.
Pinto, M., Appay, M. D., Simon-Assman, P., Chevalier, G., Dracopoli, N., Fogh, J., and Zweibaum, A. (1982) Biol. Cell *44:* 193–196.
Pinto, M., Robine-Léon, S., Appay, M. D., Kedinger, M., Triadon, N., Dussaulx, E., Lacroix, B., Simon-Assmann, P., Haffen, K., Fogh, J., and Zweibaum, A. (1983) Biol. Cell *47:* 323–330.
Pitelka, D. R., and Taggart, B. N. (1983) J. Cell Biol. *96:* 606–612.
Quaroni, A. (1985a) J. Cell Biol *100:* 1601–1610.
Quaroni, A. (1985b) J. Cell Biol. *100:* 1611–1622.
Quaroni, A. (1986) Proc. Natl. Acad. Sci. USA *83:* 1723–1727.
Rieker, J., Borysenko, C., Swanljung-Collins, H., Lamperski, B., Montibeller, J., and Collins, J. (1985) J. Cell Biol. *101:* 160a.
Rodewald, R., Newman, S. B., and Karnovsky, M. J. (1976) J. Cell Biol. *70:* 541–554.
Sandstrom, B. (1971) Cytobiologie *3:* 293–297.
Schollmeyer, J. V., Goll, D. E., Tilney, L., Mooseker, M., Robson, R., and Stromer, M. (1974) J. Cell Biol. *63:* 304a.
Stidwill, R. P. and Burgess, D. R. (1986) Develop. Biol. *114:* 381–388.
Stidwill, R. P., Wysolmerski, T., and Burgess, D. R. (1984) J. Cell Biol. *98:* 641–645.
Swanljung-Collins, H., and Collins, J. (1985) J. Cell Biol. *101:* 403a.
Thuneberg, L., and Rostgaard, J. (1969) J. Ultrastruct. Res. *29:* 578a.
Tilney, L. G., and Cardell, R. R., Jr. (1970) J. Cell Biol. *47:* 408–422.
Tilney, L. G., and Mooseker, M. (1971) Proc. Natl. Acad. Sci. USA *68:* 2611–2615.

Trier, J. S., and Madara, J. L. (1981) in *Physiology of the Gastrointestinal Tract* (L. R. Johnson, J. Christensen, M. I. Grossman, E. D. Jacobson, and S. G. Schultz, eds.), pp. 925-961. Raven Press, New York.

Verner, K., and Bretscher, A. (1983) Eur. J. Cell Biol. *29:* 187-192.

Verner, K., and Bretscher, A. (1985) J. Cell Biol. *100:* 1455-1465.

Walsh, T. P., Weber, A., Davis, K., Bonder, E., and Mooseker, M. S. (1984a) Biochemistry *23:* 6099-6102.

Walsh, T. P., Weber, A., Higgins, J. Bonder, E. M., and Mooseker, M. S. (1984b) Biochemistry *23:* 2613-2621.

Wang, Y., Bonder, E. M., Mooseker, M. S., and Taylor, D. L. (1983) Cell Motil. *3:* 151-165.

Yin, H. L., Albrecht, J. H., and Fattoum. (1981) J. Cell Biol. *91:* 901-906.

Zweibaum, A., Triadou, N., Kedinger, M., Augeron, C., Robine-Léon, S., Pinto, M., Rouseet, M., and Haffen, K. (1983) Int. J. Cancer *32:* 407-412.

Zweibaum, A., Pinto, M., Chevalier, G., Dussaulx, E., Triadon, N., Lacroix, B., Haffen, K., Brun, J.-L., and Rousett, M. (1985) J. Cell Physiol. *122:* 21-29.

THE LYSOSOMAL CONCEPT:

FROM ISOLATED PARTICLES TO ACID HYDROLASE COMPARTMENTS OF THE CELL

Phyllis M. Novikoff

DEDICATION

This review is dedicated to the loving memory of Dr. Alex Novikoff (February 28, 1913–January 9, 1987), Professor of Pathology, Albert Einstein College of Medicine, Yeshiva University.

I. INTRODUCTION

It is now more than 30 years since de Duve postulated the existence of lysosomes and showed, with his colleagues, that numerous hydrolytic enzymes were associated with these organelles (de Duve et al., 1969; de Duve and Wattiaux, 1966; de Duve, 1969, 1975). De Duve deduced, on the basis of sedimentation characteristics and latency properties, the size (0.5 μm), shape (spherical), and membrane-delimited nature (latency) of the isolated particles. Novikoff and colleagues reported the first morphological study of isolated lysosomes (Novikoff et al., 1956). Their electron microscopic examination of the particles in the "L fraction" isolated from liver homogenates confirmed these features. In addition, those morphologic observations

Advances in Cell Biology, Vol. 1, pages 59–95.

ISBN: 0-89232-792-8

revealed that the delimiting membrane was tripartite and that the lysosomal matrix contained electron-opaque particles of the size of ferritin. These ferritin particles led Novikoff et al. (Novikoff et al., 1956) to suggest that these "dense bodies" corresponded to the "pericanalicular dense bodies" seen by Rouiller in sections of rat hepatocytes (Rouiller, 1960). The development of a cytochemical procedure for acid phosphatase, the enzyme that was a marker of a unique fraction (designated L fraction for light mitochondrial fraction) obtained from homogenates of rat liver different from all other fractions, permitted a direct demonstration between the enzyme content of the isolated lysosomes and lysosomes in intact hepatocytes (Essner and Novikoff, 1961; Holt and Hicks, 1961).

These earlier investigations contributed to an understanding of the role of lysosomes in physiological and pathological processes and led to an appreciation of the heterogeneous nature of the lysosome population and to the discovery of the process of autophagy (de Duve, 1969; Novikoff, 1973; Novikoff and Shin, 1978). These studies also had a major influence on recent research on the biogenesis of lysosomes and lysosomal enzymes. Mechanisms for lysosomal enzyme synthesis, their transfer in the endoplasmic reticulum (ER), and transport to their resident organelle are being revealed. A "signal recognition protein" that mediates the unidirectional transfer of proteins into the ER lumen after their synthesis on polysomes has been characterized (Walter et al., 1984). The direction of proteins to their destinations after their translation is considered to be carried out by "sorting" signals. One signal, the mannose-6-phosphate receptor, has been identified as the signal that targets lysosomal enzymes to lysosomes (Creek and Sly, 1984; Reitman and Kornfeld, 1981).

II. GERL CONCEPT

This concept was introduced by Alex Novikoff in 1964 (Novikoff, 1964) based on studies of frozen sections of dorsal root ganglion neurons incubated for either acid phosphatase (AcPase) activity or thiamine pyrophosphatase (TPPase) activity (Novikoff and Novikoff, 1977). Distinctly different structures were stained with two different enzymes. The TPPase localizations showed curved or crescent-shaped structures; these structures appeared to form a continuous network as the focus was changed in the light microscope (Figure 1, upper). This enzyme delineated the apparatus described by Camillo Golgi (Golgi, 1898). However, with acid phosphatase localizations, using either β-glycerophosphate or cytidine monophosphate as substrate, dot-like structures and elongated irregular structures were visualized in the cytoplasm (Figure 1, lower). The dot-like spherical structures, which were lysosomes, and the elongated structures, which corresponded to portions

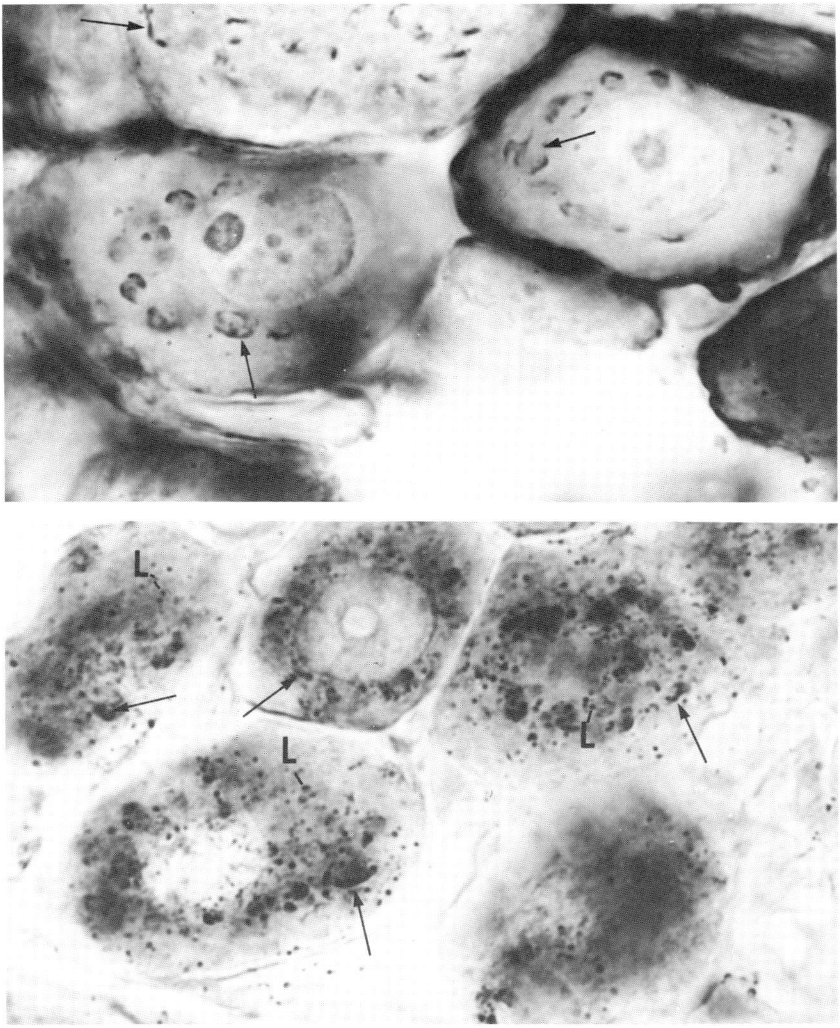

Figure 1. *Upper figure:* Rat dorsal root ganglia incubated for TPPase activity. Arrows indicate the horseshoe-shaped portions of the Golgi apparatus. ×2000.

Lower figure: Rat dorsal root ganglia incubated for AcPase activity. Lysosomes (dot-like structures) and GERL (curved structures) are evident. ×2000. (From Novikoff and Novikoff, 1977, with permission.)

of GERL, were in close association with each other and occupied an area bordered by the Golgi apparatus.

Novikoff's original definition (Novikoff, 1964) of the acronym GERL was as follows:

> Between the cell nucleus and the innermost of the three *G*olgi saccules there is a region of smooth *ER* with a complex, specialized structure in which localized accumulations of ferritin-like grains and continuities with "coated vesicles" are common. The images suggest that both dense bodies (lysosomes) and "coated vesicles" arise from it. Its continuity with ribosome-studded ER has been observed but not with Golgi saccules. . . . The special region of ER is referred to as GERL to suggest that it is intimately related to the Golgi saccule (*G*), that it is part of the *ER* and that it forms lysosomes (*L*).

Morphological and cytochemical studies at the electron microscope level on neurons (Figure 2) and many other cell types supported the notion that a structure was found in the cytoplasm which was distinct from the Golgi apparatus. In all cell types where GERL is present, the essential characteristic is the presence of acid hydrolases in smooth membrane cisternae. The smooth membrane cisternae of GERL resemble the smooth ER in being

---

Figure 2. Sections of neurons from rat dorsal root ganglia. The distinctive morphology and enzymatic activity of GERL are illustrated in these figures.

Upper figure: The Golgi apparatus consists of several cisternae, which are parallel to each other and separated from each other by a fixed distance. The group of cisternae form a stack (*G*). GERL appears as an irregular network of cisternae and tubules, which have coated regions in areas. Portions of GERL are indicated by arrows. Small vesicles are found between the Golgi stack and GERL cisternae. Vesicles are not present between the individual Golgi cisternae. The Golgi apparatus appears as a well-ordered array of cisternae. GERL has an irregular arranged membranous system—cisterna, coated areas (*C*), and tubules. Continuity of GERL with rough endoplasmic reticulum (*RER*) is evident at the crossed arrow. Numerous free ribosomes (*R*) are close to GERL structures. ×40,000.

Lower figure: Incubated for AcPase activity. Reaction product is restricted to GERL (*arrows*), lysosomes, and coated vesicles (*C*). The irregular anastomosing network is more sharply delineated when tissue is incubated for AcPase activity. Note that the Golgi stack and other organelles show no AcPase activity. Rough endoplasmic reticulum (RER) is directed toward GERL. Huge numbers of ribosomes (*R*), apparently free in the cytoplasm, are close to the GERL region. ×25,000.

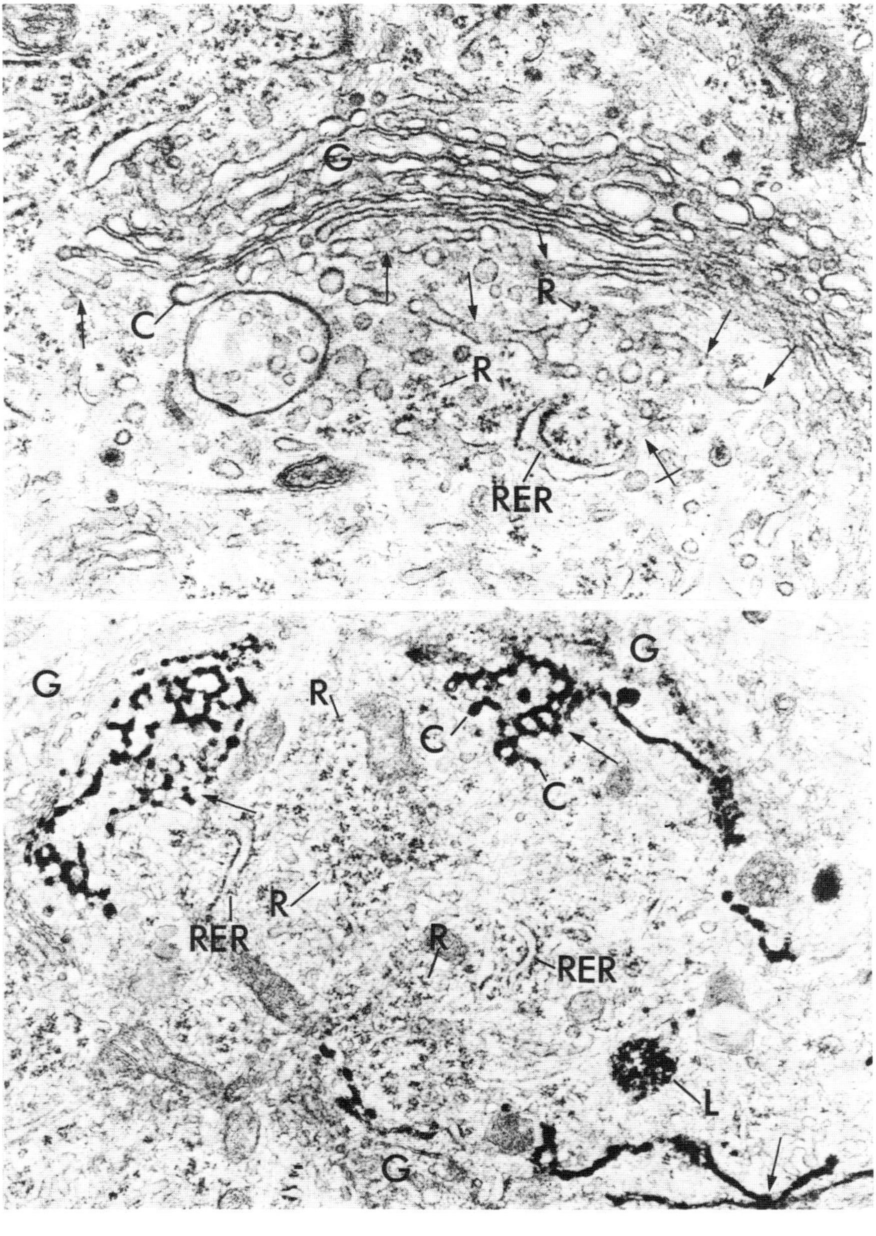

Figure 2

devoid of ribosomes on its surface. Direct continuities of GERL and the ER were observed, but not between GERL and the Golgi apparatus. The close spatial relation between GERL and the Golgi apparatus indicates a probable functional interplay.

For a long time the only acid hydrolase demonstrated in GERL and lysosomes was acid phosphatase; however, our laboratory and others have localized other lysosomal enzymes (e.g., aryl sulfatase, E-600-resistant thiolacetic acid esterase, β-galactosidase, β-glucuronidase, acid mannosidase, trimetaphosphatase, cathepsin D, acid α-glucosidase) in GERL and lysosomes in a variety of cell types (Bentfeld and Bainton, 1975; Decker, 1974; Novikoff et al., 1983; Novikoff et al., 1985; Oliver, 1980; Decker et al., 1980; Iwamasa et al., 1982; Holtzman and Diminitz, 1968; Pino et al., 1981; Pelletier and Novikoff, 1972; Boutry and Novikoff, 1975; Essner and Oliver, 1974; Novikoff et al., 1979; Novikoff et al., 1977; Novikoff et al., 1980, 1971, 1975; Novikoff and Yam, 1978a; this is a partial compilation). Thus, with the *in situ* demonstration of a number of other lysosomal enzymes in GERL, the original idea that lysosomes (coated vesicles, autophagic vacuoles, residual or dense bodies) arise from GERL was given much support.

The novel implication raised by the GERL concept is that the transport of lysosomal enzymes into lysosomes, after their synthesis, may not involve the Golgi apparatus. Lysosomal enzymes may be transported directly from the ER to lysosomes via GERL. If so, this pathway would differ from the one used for the transport of secretory proteins. The lysosomal enzymes then would bypass the Golgi stack. If lysosomal enzymes (most of which are glycoproteins) bypass all or part of the Golgi stack, then there must exist in the cytoplasm additional intracellular compartments for modification of the enzymes after synthesis (e.g., partial glycosylation of enzymes occurs in the Golgi apparatus). Candidates for these sites could be the ER or GERL, or intracellular sites yet to be identified.

III. STRUCTURE AND FUNCTIONS OF GERL

This review will focus on two cell types, the neuron and the hepatocyte, to illustrate the morphological and enzymatic properties of GERL (Figures 2-6).

In the neurons and hepatocytes, GERL consists of cisternae and tubules that anastomose, forming a complex reticulum network (Novikoff et al., 1971, 1975; Novikoff and Yam, 1978a,b). The anastomosing tubules sometimes appear like cisternae, particularly if they are distended. It has not been determined either how extensive the anastomosing network of tubules and cisternae is or how far this system extends away from the transmost

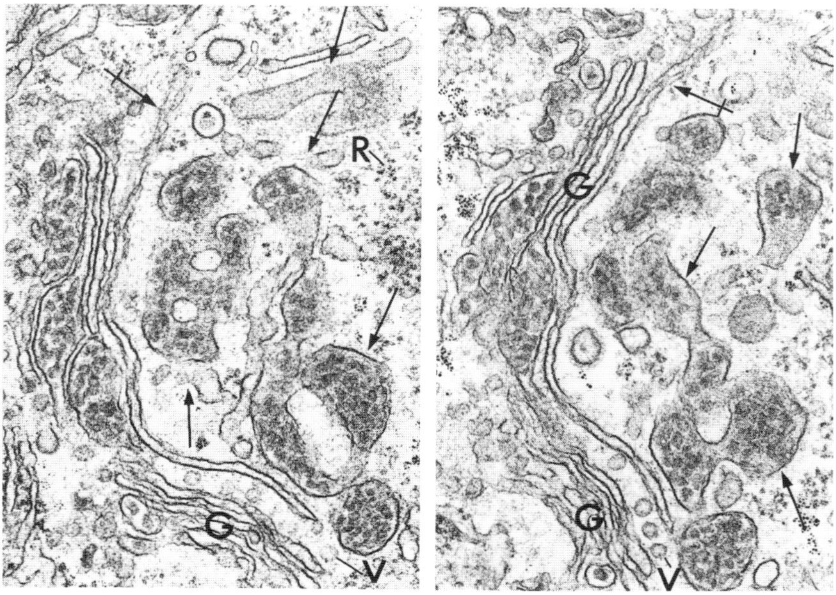

Figure 3. Serial sections of hepatocytes from an untreated rat liver. The anastomosing nature of GERL (*arrows*) is distinct from the Golgi stack (*G*). Electron-dense material and lipoprotein particles are evident within GERL cisternae (*arrows*); these particles are more electron-opaque than those in the Golgi cisternae. A portion of a flattened cisterna of GERL (*crossed arrows*) lies close to the stack; other portions are separated from the stack by a variable distance. Vesicles (*V*) are located between the stack and GERL. Numerous ribosomes (*R*) lie free in the cytoplasm close to GERL. *Left and right*, ×43,000. (Novikoff and Yam, 1978a, with permission.)

Golgi cisterna. Acid phosphatase-reactive cisternae are seen sometimes lying close to the *trans*-Golgi cisterna, whereas at other times reactive cisternae and tubules extend over a wide area removed from the *trans*-Golgi cisterna (Figure 2, lower; Figure 4, left). In a study on neurons in which sections 10–20 times thicker than the usual thin Epon sections (and incubated for acid phosphatase activity) were examined with the electron microscope, an extensive irregular tubular network was revealed; this network occupied a large area that was *trans* to the Golgi apparatus [see Figures 17–29 in Novikoff et al., 1971].

Numerous free ribosomes are seen in the GERL region (Figures 2 and 3). Continuities between rough ER and GERL are found in the immediate

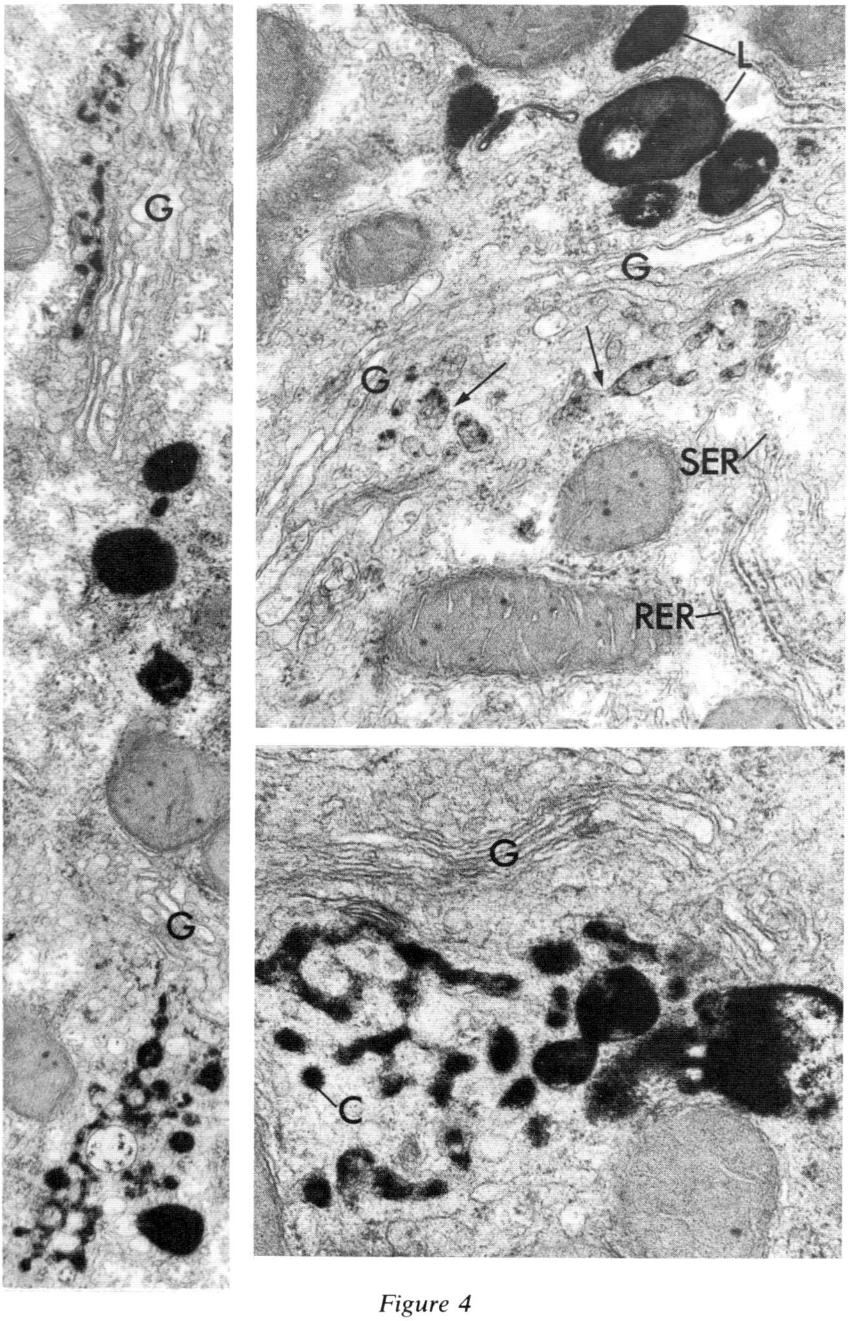

Figure 4

vicinity of GERL (Figures 2; 3, left) as well as in regions distant from GERL (Figure 6).

Other characteristics of GERL include coated areas or coated vesicles attached to membrane of GERL (Figures 2; 4, lower right; 6). Often electron-opaque material is found within the cisternal space, and occasionally a clear region is seen between the delimiting membrane and the finely granular matrix (Figures 3 and 6). The distended regions of the cisternal and tubular areas may be focal regions of lysosome formation (Figures 3; 4, left, lower right). It is not known if these regions are randomly distributed or if lysosomal formation takes place in special subcompartments in the GERL reticulum. The more compressed or flatter cisternal portion of GERL sometimes lies close to and parallel to the *trans*-Golgi stack (Figures 2 and 3); however, often this cisternae extends beyond the Golgi stack while retaining its parallel orientation (Figures 2 and 3). Frequently, vesicles are seen between the cisternae and the Golgi stack (Figure 3). In hepatocytes, lipoprotein particles of unknown composition are present in GERL (Figures 3; 4, upper right; 5; 6). Apoprotein B has been localized within structures resembling GERL, suggesting that these particles are very-low-density lipoprotein particles (VLDL) or low-density lipoprotein particles (Alexander et al., 1976). Based on electron microscopic images, we have suggested that these particles and other macromolecules gain access to GERL by direct continuities to the ER (Novikoff and Yam, 1978a). An alternate pathway for transport of macromolecules to GERL may involve smooth tubules of ER which bypass the Golgi stack. Figure 5, lower, illustrates ER tubules in an apparent interruption of the Golgi stacks. In reality, the Golgi

---

Figure 4. This figure emphasizes the enzymatic and morphological differences between GERL and the Golgi stack. Sections of rat liver were incubated for AcPase activity. Reaction product delineates the extensive anastomosing tubular and cisternal structures of GERL. Reaction product is not found in the Golgi stack. Portions of GERL are seen close to the Golgi stack; other portions are far removed from the stack (*left figure*). Acid phosphatase-positive lysosomes (*L*) are evident at *cis* (*upper right*) and *trans* (*lower right*) aspects of the Golgi stack. Lipoprotein particles and AcPase reaction product (arrows) are seen in GERL cisternae. A parallel array of rough endoplasmic reticulum (*RER*), continuous with smooth endoplasmic reticulum (*SER*), is directed toward GERL. Dilatations are found in regions of the anastomosing tubules of GERL (*left and lower right*) as well as coated vesicles (C). *Left*, ×25,000; *upper right*, ×30,000; *lower right*, ×52,000. [From Novikoff and Yam, 1978b, with permission.]

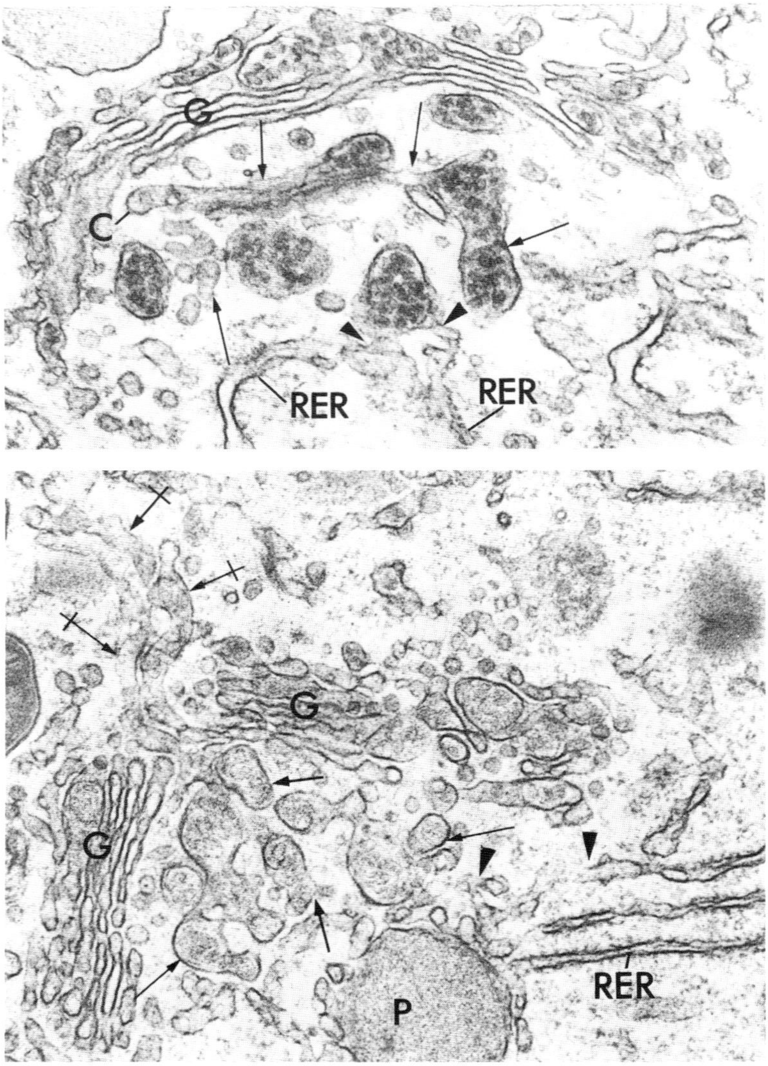

Figure 5. Upper figure: Portion of a hepatocyte from the liver of an untreated rat. GERL (*arrows*) is continuous with rough ER (*RER*) at the arrowheads. Electron-dense lipoprotein particles are seen within cisternae of GERL. These particles are denser than those within the Golgi cisternae. Portions of GERL are separated from the stack by a fixed distance, whereas the individual Golgi cisternae are separated by a fixed distance. ×45,000. (From Novikoff and Yam, 1978a, with permission.)

stacks are continuous at another plane of the section. In addition, circulating exogenous lipoprotein particles may be endocytosed and transferred to GERL. These particles may undergo modifications in their protein, carbohydrate, or lipid moieties that could lead to partial or complete catabolism, or the particles could be altered under the influence of hydrolases as an additional processing even before secretion.

Although extensive functional studies have not been performed in neurons, GERL appears to be involved in the formation of lysosomes (Novikoff et al., 1971), in ribonucleoprotein turnover in normal and chromatolytic neurons (Holtzman et al., 1967) and in the processing of internalized neuronal plasma membranes (Gonatas et al., 1977; Kim et al., 1978).

In 1971, we raised the possibility that GERL might also function in the maturation of glycoproteins in neurons (Novikoff et al., 1971). We were intrigued by the frequent observation that a portion of GERL was in close apposition to the innermost Golgi cisterna and that tubules of GERL extended into the polygonal compartments of the innermost Golgi cisterna (the polygonal compartments were revealed by examining 0.5-μm sections incubated for TPPase activity) (Novikoff et al., 1971). This observation, combined with (1) the presence of a nucleoside diphosphatase (TPPase) in the *trans*-Golgi element or cisternae, (2) the presence of cytidine monophosphatase in GERL and lysosomes, and (3) reports that glycosyltransferases were considered to be localized in "Golgi-enriched fractions" (Wagner and Cynkin, 1969; Fleischer et al., 1969; Morré et al., 1969; Schachter et al., 1970), led us to reason that there must be a potential physiological significance for both the enzymatic properties and the close topological relation between GERL and the Golgi apparatus. It was known that galactosyltransferase and sialyetransferase generated nucleoside diphosphate (uridine diphosphate) and cytidine monophosphate, respectively (Schachter et al., 1970). However, at that time there were no *in situ* methods to localize

---

Figure 5—continued

Lower figure: Portion of a centrolobular hepatocyte from the liver of a diabetic rat bearing transplanted pancreatic islet. Smooth ER tubules containing lipoprotein particles (*crossed arrows*) are seen in a "passageway" or interruption of the Golgi stack (*G*). Lipoprotein particles are seen within anastomosing cisternae of GERL (*arrows*). Continuities of rough ER (*RER*), smooth ER (*SER*), and GERL are evident at the arrowheads. Macromolecules could be transported into GERL via smooth ER tubules or continuities with the ER, thereby effectively bypassing the Golgi stack. ×50,000. (From Eder et al., 1979, with permission.)

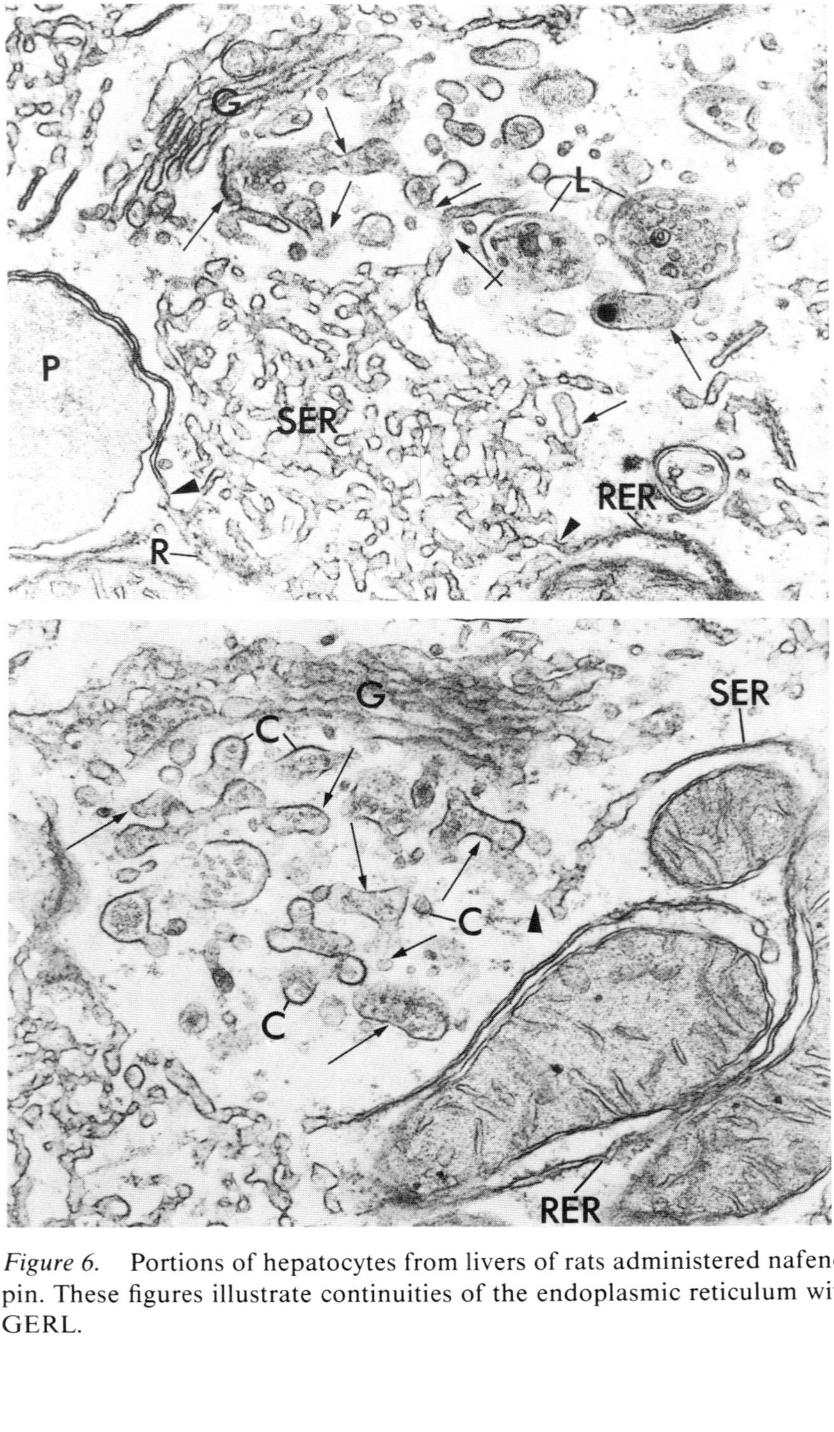

Figure 6. Portions of hepatocytes from livers of rats administered nafenopin. These figures illustrate continuities of the endoplasmic reticulum with GERL.

glycosyltransferase activities. Recently galactosyltransferase was co-localized with the TPPase-positive *trans*-Golgi cisternae (Roth and Berger, 1982), and sialyetransferase with the cytidine monophosphatase (AcPase)-positive GERL cisternae (Roth et al., 1985). Electron microscopic autoradiographic studies demonstrated that the addition of sialic acid to glycoproteins occurred in GERL structures identified by AcPase cytochemistry (Bennett and O'Shaughnessy, 1981). The functions of the enzyme cytochemical markers have since been more biochemically defined. It has been shown that (1) nucleoside diphosphatase is capable of rapidly hydrolyzing uridine diphosphate that is released when galactose is transferred from uridine diphosphate-galactose to the nascent glycoprotein chain, and (2) cytidine phosphatase hydrolyzes cytidine monophosphate released after sialic acid is transferred from cytidine monophosphate-sialic acid to the nascent glycoprotein chain (Kuhn and White, 1977). The enzyme cytochemical markers were not only distinguishing different structures but were also revealing distinct functions that are currently under study.

In hepatocytes, GERL may be an additional site where secretory proteins are processed and packaged into vacuoles. Vacuoles containing both acid phosphatase activity and lipoprotein-like particles are present in both the Golgi zone (Figure 4, upper right) and at the sinusoidal aspect of the hepatocyte. We have not determined whether these vacuoles secrete their content into the space of Disse by exocytosis. It is also possible that GERL may function in a manner analogous to that described in the pituitary cells in which excess secretory product is degraded in a process called *crinophagy* (Farquhar, 1969). GERL may not only be the site of modifying secretory proteins but may also act in controlling the level and rate of secretion. The role of GERL in the processing of secretory material and in the involvement with the formation of secretory granules has been investigated in a number of other secretory cells (e.g., adrenal medulla, salivary gland, anterior pituitary, submandibular gland, lacrimal gland, parotid gland, and exocrine

---

Figure 6—continued

Upper figure: Direct continuity of GERL (*arrows*) with smooth ER (*SER*) is evident at the crossed arrow. This SER is, in turn, continuous with rough ER (*RER*) (*arrowheads*). Also labeled are Golgi stack (*G*), lysosomes (*L*), and peroxisome (*P*). ×30,000.

Lower figure: GERL (*arrows*) is in direct continuity with an elongated cisternae of smooth ER (*SER*), which is connected to rough ER (*RER*). Coated regions (*C*) are evident on GERL cisternae. The individual cisterna of the Golgi stack (*G*) is not distinct because the stack is sectioned tangentially. ×30,000.

pancreas) (Holtzman and Dominitz, 1968; Pelletier and Novikoff, 1972; Novikoff et al., 1977; Hand and Oliver, 1977a,b).

In addition to the function of GERL in the modifications of secretory lipoprotein proteins (VLDL particles), GERL has also been shown to participate in the metabolism of lipids that accumulate in the cytoplasm (either as nonmembrane-delimited neutral lipid cytosolic spheres or as lipid spheres within vesiculated endoplasmic reticulum) (Novikoff et al., 1974; Novikoff and Edelstein, 1977; Novikoff, 1977; Nehemiah and Novikoff, 1974). Massive accumulations of lipids can be induced by dietary means to the extent that a grossly visible fatty liver is produced, and this fatty liver can subsequently be reversed, again by dietary means. Whether the lipid accumulates in the cytosol or in the endoplasmic reticulum, the excess lipid is metabolized by a coordinated interplay of several organelles, including the endoplasmic reticulum, peroxisomes, and microperoxisomes (Novikoff et al., 1974; Novikoff and Edelstein, 1977; Novikoff and Novikoff, 1972; Novikoff et al., 1973a,b). Under these conditions there is a dramatic hypertrophy of GERL and an increase in the number of residual bodies; lipid inclusions are seen in both GERL and residual bodies (Novikoff and Yam, 1978). Probably the most extensive development of GERL and large residual bodies that accumulate lipid-like inclusions is found in the hepatocytes of the beige mutant mouse, which is considered to be a homologue of the Chédiak-Higashi syndrome in man; the lipid accumulations are thought to occur because of a disturbance of lipid or lipoprotein metabolism (Oliver and Essner, 1973).

Many proteins are synthesized as precursor molecules (e.g., pro-albumin, pro-insulin, and lysosomal enzymes). Precursor molecules are modified by proteolytic processing intracellularly to their mature forms. Although the intracellular site in which proteolytic cleavage occurs has not been identified, there is some biochemical evidence that the precursor form of the lysosomal enzyme, α-mannosidase, is modified in a membrane fraction that has the properties of lysosomes (Wood and Kaplan, 1985). It was also reported that the modification of β-galactosidase and cathepsin D into mature forms occurs in structures that may be lysosomal in nature (Erikson et al., 1981; Rosenfield et al., 1982; Sabatini et al., 1982; Gieselmann et al., 1983). It is possible that GERL or other components of the lysosomal system may be the intracellular sites of the modification of precursor molecules.

Although the functions of GERL may vary depending on the cell type, it is likely that in those cell types in which it is present the common functions are concentration and possible sorting of lysosomal enzymes, degradation of endogenous and exogenous macromolecules, and formation of lysosomes. It is not known whether different functions occur in different parts of GERL.

Thus, glycosylation of lysosomal enzymes may take place in the cisternal parts that are in close apposition to the *trans*-Golgi cisternae. In contrast, the formation of lysosomes may occur in a part of GERL distant from the *trans*-Golgi cisternae. Acid hydrolases are found throughout the system. This may indicate that modifications (e.g., hydrolysis or degradation) are not confined to a subcompartment of GERL but, rather, occur throughout.

Although numerous functions have been attributed to GERL—e.g., formation of autophagic vacuoles (Novikoff, 1973), formation of the central cell vacuole in plants (Marty, 1978), formation of secretory granules in pancreas (Novikoff et al., 1977) and in lacrimal and other exocrine glands (Hand and Oliver, 1977b), transport of lipids in the liver (Novikoff and Yam, 1978b), locus of endocytosed colloidal silver in macrophages (Essner and Haimes, 1977) and receptor bound ligands such as lectins and cholera toxin (Gonatas et al., 1977; Joseph et al., 1978) endocytosed ferritin in anterior pituitary cells (Farquhar, 1978), and reservoir of membranes in cultured fibroblasts (Garvin et al., 1981)—the isolation and partial characterization of GERL has only been reported in human cultures fibroblasts (Rome et al., 1979).

Are GERL cisternae always found parallel to the Golgi apparatus? Since we have not determined the three-dimensional structure of either GERL or the Golgi apparatus, it is not fully possible to answer this question. The Golgi stack can be sectioned without sectioning the GERL structures, and GERL cisternae can be sectioned without passing through the Golgi stack. Since one can cut the Golgi stack without seeing GERL, it is possible that GERL is not in register with the width of the stack. This out-of-register topography between these two organelles could explain, in part, the apparent absence of recognizable GERL structures where the Golgi stack is cut perpendicularly.

We have also seen images in which acid hydrolase-rich structures are present on both the cis and trans side of the Golgi stack (Figure 4, upper right) and in regions between separate Golgi stacks. As was stated earlier, the Golgi stacks are probably not discrete units but are connected at another level of the section. These variable morphological appearances of GERL indicate that this organelle is not only a very dynamic organelle but also a very complicated structure. A three-dimensional study of GERL in regenerating liver by serial sectioning indicates its complex nature (Jaeken et al., 1978). This study also confirmed our observations that continuities exist between the ER and GERL. High-voltage electron microscopic studies of tissue incubated for a number of acid hydrolases are in progress. These studies are expected to aid us in determining the three-dimensional reconstruction of GERL and also of the lysosomal system (discussed below).

IV. DISTINCTIONS BETWEEN GERL AND GOLGI APPARATUS

For a long time there had been considerable debate about the existence of the Golgi apparatus (Lacey and Challice, 1957). Two important developments occurred demonstrating the Golgi apparatus to be a real structure in the cytoplasm, not an artifact of fixation: one was publication of electron microscopic images of neurons showing stacks of parallel smooth membrane structures (Palay, 1976; Palay & Palade, 1955); the other was the development of an enzymatic cytochemical procedure (Novikoff and Goldfischer, 1961) that gave images identical to those seen by Camillo Golgi using osmium staining (Golgi, 1898).

The British cytologist John R. Baker had long contested the view that the classical Golgi apparatus as shown by Camillo Golgi in neurons corresponded to the "lamellar-vacuolar fields" seen in electron micrographs by numerous investigators (Dalton and Felix, 1956). Dr. Baker showed great integrity in changing his long-held view regarding the Golgi apparatus. He wrote (Baker, 1963):

> My own acceptance of the classical view has been brought about by evidence from four different sources. First, I have myself noticed that in sympathetic neurons the Nissl substance tends to be peripheral, while the Golgi apparatus is arranged round the nucleus. Secondly, Thomas (1963)[1] has shown that if a silver preparation of a vertebrate neuron is photographed to show the exact arrangement of the Golgi apparatus, and if the coverslip is then removed and the preparation treated by a routine method for the Nissl bodies, the distribution of the two bodies is different. He has performed variations of this experiment, in one of which I myself made a careful drawing of a particular cell before the coverslip was removed, and examined the same cell again, with the drawing beside me, when the Nissl technique had been applied. Thirdly, I was shown a very large and remarkable electron micrograph of a Purkinje cell of the rat, made by Dr. S. L. Palay, in which the lamellar-vacuolar fields had been everywhere marked in red: the arrangement and amount fully justified the conclusion that these fields might represent the Golgi apparatus.
>
> The episode that finally changed my opinion was the publication of a paper by Novikoff and Goldfischer (1961)[2] and the microscopical slides given to me by Dr. Novikoff on the occasion of his visit to Oxford.
>
> Dr. Novikoff and his associates have made a special study of the distribution of phosphatases in cells, and have developed new techniques for the cytochemical localization of some of these enzymes. The enzyme that is of particular importance from the point of view of the Golgi controversy is thiamine diphosphatase.... It is thus possible to use the thiamine diphosphatase reaction as what Novikoff calls a "marker" for the Golgi apparatus (Novikoff, 1962).[3]

With time it was shown that almost all cell types had a Golgi apparatus and that the innermost cisternae were reactive for TPPase and the outermost for osmium deposition.

It was pointed out to us that thiamine pyrophosphatase activity was not found in all the cisternae of the apparatus but was confined to only one element of the Golgi apparatus, the trans element (Cheetham et al., 1971). The enzyme cytochemical method gave the same image as the classical technique because the osmium-reducing outer cisternae and the thiamine pyrophosphatase reactive inner cisternae are parallel to each other and remain so along the entire length of the extensive apparatus.

Even though nucleoside diphosphatase (TPPase) is also present in the endoplasmic reticulum, bile canaliculus, and plasma membrane of other cell types, the application of the enzyme cytochemical procedure to "mark" the Golgi apparatus continues to be valid at both the light and electron microscope levels.

In most cell types the Golgi apparatus consists of a variable number of cisternae or elements bounded by smooth membranes. In the neurons and hepatocytes three to six cisternae are present (Figures 2–6). A group of cisternae is referred to as a Golgi stack. The cisternae always remain parallel to each other and are separated from each other by a relatively constant distance. The cisternae are held in a fixed position in relation to each other by unknown mechanisms. Occasionally a dense material is seen between cisternae; it has been suggested that this intercisternal material has adhesive properties and holds the cisternae together in a stack (Mollenhauer and Morré, 1966). The extensive Golgi apparatus undulates and twists throughout the cytoplasm. Its three-dimensional appearance is difficult to appreciate in thin sections where portions of the apparatus are seen as separate stacks. These separate units are continuous and form one Golgi apparatus. Often, but not always, the stacks are curved or crescent shaped. The cis and trans aspects are not always easily recognized because of the twisting and turning of the apparatus.

Whereas the morphology of the Golgi apparatus is easily recognized because its structure is well defined, GERL has a more varied appearance. Depending on the cell type, it consists of smooth-surfaced flattened cisternae, distended cisternae, and anastomosing tubules. All these structures are interconnected; however, it is rare to find all portions of GERL in a given section. Coated regions are found on its surface and coated vesicles are attached to its surface. The coated vesicles in the GERL region are smaller (~50 nm) than the coated vesicles at the cell surface and, like all other lysosomes, contain acid phosphatase activity (Novikoff et al., 1980). The coated vesicles are interpreted as forming from GERL. The content of GERL also varies with the cell type. Secretory material, dense material of unknown nature, ferritin-like particles, and lipoprotein particles have been found within GERL. A clear zone or halo is often found between the matrix and the delimiting membrane. An important feature that has not received sufficient attention is the numerous free ribosomes in the GERL region.

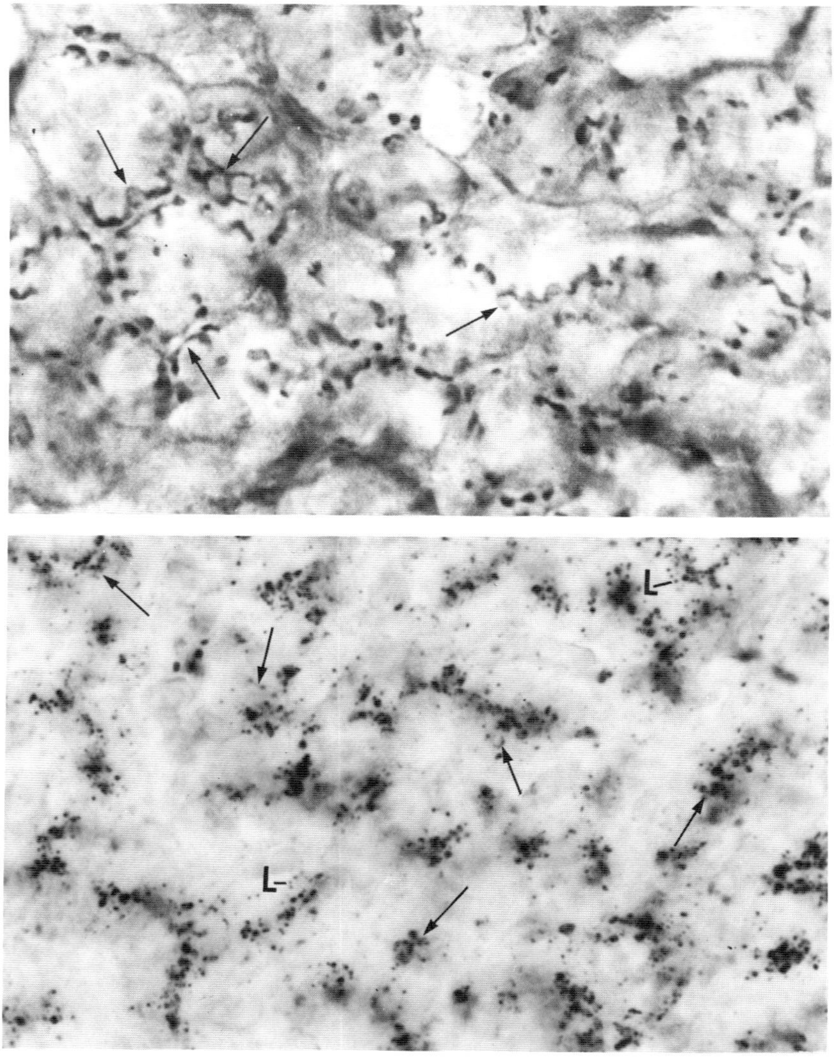

Figure 7. Nonfrozen (Vibratome) sections of untreated rat liver.

Upper figure: Immunocytochemical localization of α-D-mannosidase II. This enzyme is localized in the Golgi apparatus of the hepatocytes. The widespread nature of the organelle and its tortuous, convoluted appearance are evident. Portions of the apparatus are seen along the bile canaliculus, near the nucleus, and elsewhere in the cytoplasm (*arrows*). ×600. (From Novikoff et al., 1985a, with permission.)

These were originally interpreted as resulting from the "degranulation" of ribosomes from the rough ER, thereby forming the smooth membrane system of GERL (Novikoff et al., 1977). It is also possible that synthesis of specific proteins (membrane or lysosomal proteins) occurs near the GERL region. *In situ* localization of specific mRNAs by cDNA probes (which are being presently developed) may reveal synthetic events in the GERL region. Frequently, parallel arrays of rough ER are directed toward GERL (Figure 4, upper right). In some cell types, GERL may appear as a single cisterna that lies close to the innermost Golgi cisterna. This proximity makes it difficult sometimes to distinguish GERL from Golgi cisternae on morphological criteria alone; cytochemistry permits this distinction since the trans-Golgi cisternae shows TPPase activity (see Figure 10, upper) and GERL, acid phosphatase activity (Figure 4). Our view is that these two structures are clearly different in cytochemical, morphological, and functional properties, so they should be considered as two distinct organelles. Recent studies on the fate of endocytosed molecules and on exocytosis of viral proteins show that these materials enter AcPase-positive (GERL) cisternae, and not Golgi cisternae (Stockert, 1980; Reston et al., 1980; Dunn et al., 1986; Griffiths et al., 1985). Other designations are being used (e.g., *trans*-Golgi reticulum) to describe structures that were originally identified by Alex Novikoff in 1964. GERL has been shown to have a greater plasticity than the Golgi apparatus in a number of experimental conditions. A few situations are cited in which GERL is enlarged without affecting the Golgi apparatus (i.e., the Golgi apparatus retains its stack-life morphological appearance). This occurs in hepatocytes of livers of rats treated with swainsonine (Novikoff et al., 1985a) in rats given nafenopin by gavage (Novikoff et al., 1975), in rats given clofibrate to reverse an orotic acid-induced fatty liver (Novikoff et al., 1974; Novikoff and Edelstein, 1977), in hepatocytes from regenerating liver (unpublished observations), or in hepatocytes from the beige mouse, which has a genetic disorder for lipid metabolism (Essner and Oliver, 1974).

Probably the most interesting example of the difference between GERL and the Golgi apparatus occurs in mitosis (Figures 7–9). During mitosis

---

Figure 7—continued

Lower figure: Cytochemical localization of acid phosphatase using cytidine monophosphate as substrate. Lysosomes (*L*) (spherical structures) and GERL (*arrows*) (curved structures) are stained. AcPase-positive structures are concentrated along the bile canaliculus; they are also present elsewhere in the cytoplasm. Comparison of AcPase and of α-D-mannosidase distributions show that there is no overlap in the staining pattern between these enzymes. ×600.

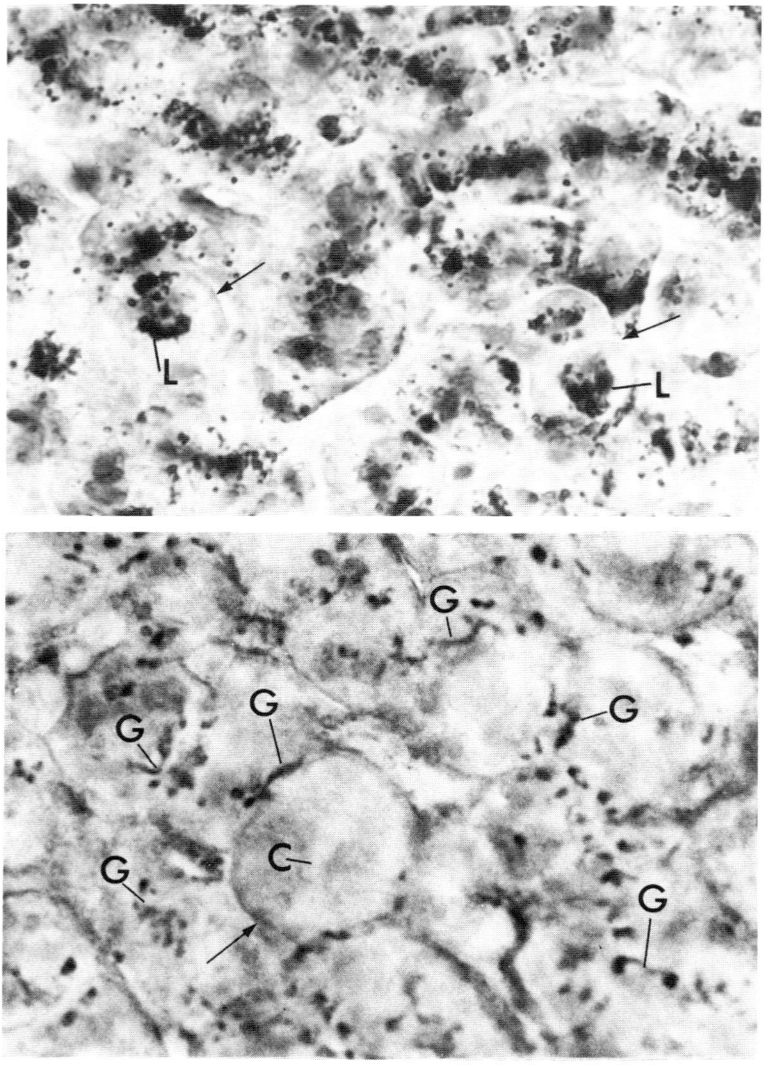

Figure 8. Sections of liver from a rat after swainsonine ingestion.

Upper figure: Frozen section incubated for AcPase activity and counterstained with Toluidine Blue. Lysosomes (*L*) are concentrated at the mitotic poles in the dividing hepatocytes (*arrows*). Lysosomes (dot-like structures) are also found along the bile canaliculus and elsewhere in the cytoplasm. ×800.

the Golgi apparatus virtually disappears, as judged by light microscopic preparations of TPPase (Robbins et al., 1964) and α-D-mannosidase activities (Novikoff et al., 1983). In contrast, GERL and lysosomes not only do not disappear but may even hypertrophy (Novikoff et al., 1985a; Mori et al., 1982). Figure 8, lower, shows the absence of α-D-mannosidase activity, and Figure 8, upper, shows the presence of AcPase activity at the mitotic poles of dividing cells. When mitotic cells are examined at the electron microscope level, the Golgi stacks are not evident (Figure 9); however, a remnant Golgi cisterna was detected in TPPase-incubated sections [see Figure 8 in Novikoff et al., 1985a]. Figure 9 shows the presence of GERL and lysosomes in hepatocytes in mitosis; Golgi stacks are not seen in these figures.

It is becoming increasingly evident that GERL is a highly dynamic structure which is capable of rapidly changing its size in response to different situations.

As stated earlier, cytochemical differences exist between the Golgi stack, and GERL and lysosomes. The cytochemical markers for the Golgi stack include osmium reduction (Cheetham et al., 1971), thiamine pyrophosphatase (Novikoff and Goldfischer, 1961; Novikoff, 1985), α-D-mannosidase (Novikoff et al., 1983), galactosyltransferase (Roth and Berger, 1982), nicotinamide adenine dinucleotide phosphatase (Smith, 1980), and *N*-acetylglucosamine transferase I (Dunphy et al., 1985). GERL has been shown to contain a number of acid hydrolases including acid phosphatase, acid mannosidase, β-galactosidase, β-glucuronidase, aryl sulfatase, E-600-resistant esterase, trimetaphosphatase (Novikoff, 1973; Bentfeld and Bainton, 1975; Decker, 1984; Novikoff et al., 1983, 1985a; Oliver, 1980; Decker et al., 1980; Iwamasa et al., 1982), and so far one glycosyltransferase (sialyetransferase) (Roth et al., 1985). Under the conditions we employ in our laboratory, the Golgi apparatus of neurons and hepatocytes (and generally for most cell types) does not show the enzymatic properties of GERL and lysosomes. Likewise, GERL and lysosomes do not show the properties of the Golgi apparatus. Figures 1 and 7 illustrate the distribution of Golgi enzymes (TPPase and α-D-mannosidase II) in the Golgi apparatus and lysosomal enzyme (acid phosphatase) in GERL and lysosomes. These

---

Figure 8—continued

Lower figure: Nonfrozen section incubated for α-D-mannosidase activity. The Golgi apparatus (*G*) is not evident in the hepatocyte in mitosis (*arrow*) and shows no mannosidase activity. In contrast, mannosidase activity is evident in the Golgi apparatus of all nondividing hepatocytes. Chromosomes (*C*) (unstained) occupy the clear area in the mitotic hepatocyte. ×1000. (From Novikoff et al., 1985a, with permission.)

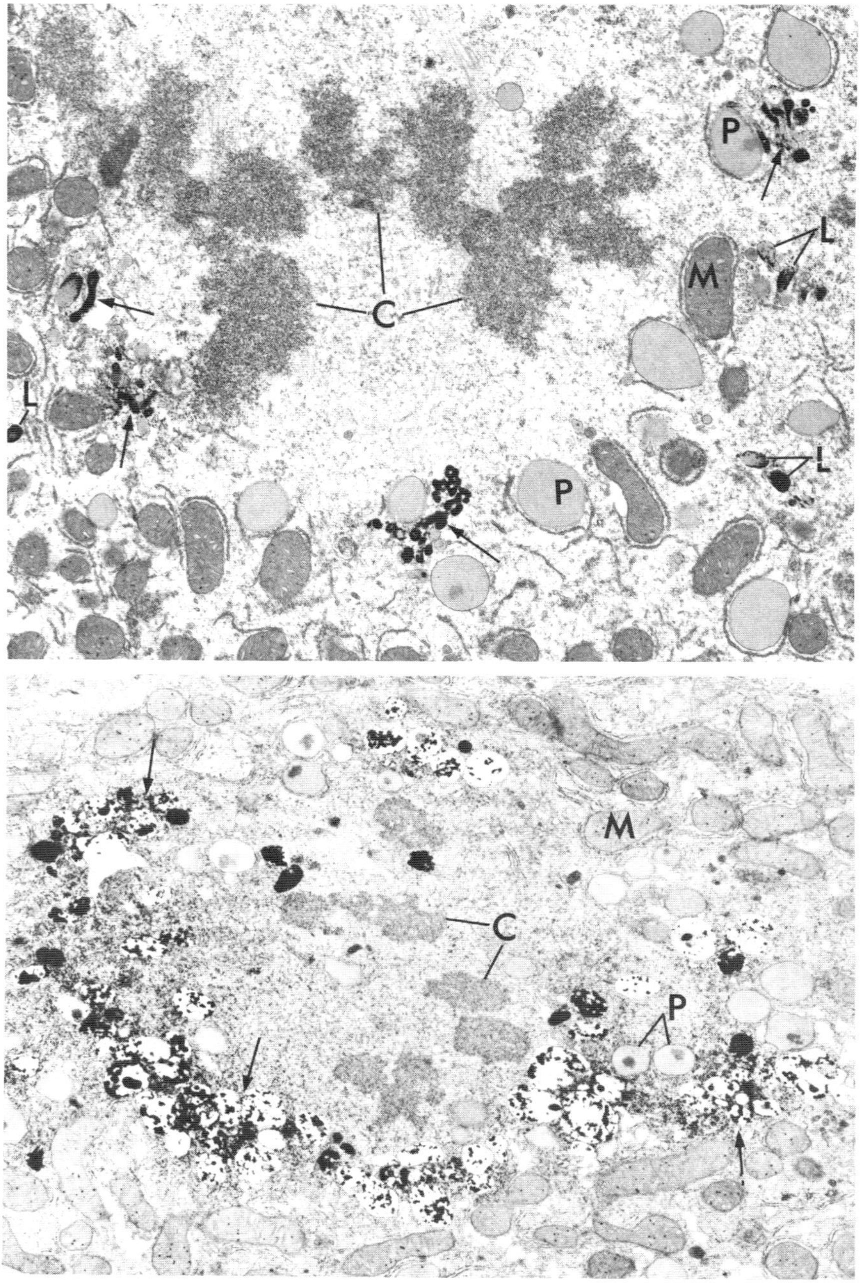

Figure 9

light microscopic studies show images that are dramatically different and without superimposition.

Recently, we have localized the glycoprotein-processing enzyme α-D-mannosidase in the Golgi apparatus using immunocytochemistry: α-D-mannosidase II, the enzyme that cleaves the final two mannosyl residues from carbohydrate intermediates, was found in all the cisternae of the entire Golgi apparatus (Figure 10, lower). Based on numerous biochemical studies, it was proposed that the processing enzymes act in a sequential fashion; that is, first there was a trimming process, with the removal of three terminal glucoses and also six of nine mannose residues, and then there was an elongation process which added *N*-acetylglucose, galactose, and sialic acid to the residual oligosaccharide chain. It was expected that the subcellular distribution of the processing enzymes would reflect their biochemical actions, that is, that each glycoprotein processing enzyme would be spatially separated in a distinct compartment or subcompartment of the Golgi apparatus. We had expected to find mannosidase II localized in the outer cisternae of the Golgi apparatus since this is an early-acting trimming enzyme of the glycoprotein biosynthetic pathway in rat hepatocytes. However, the enzyme was not found in a single Golgi cisterna, but was localized in all the cisternae. In exocrine pancreatic cells and anterior pituitary cells, α-D-mannosidase II was also found in more than one Golgi element, although not in all (Farquhar, 1985). The multiple Golgi subcompartment localization for mannosidase II differed from the other Golgi markers (e.g., TPPase; see Figure 10, upper).

Before the development of immunocytochemical procedures, the ability to demonstrate enzymes in the Golgi apparatus *in situ* was limited to the TPPase method (Novikoff and Goldfischer, 1961).

Figure 9. Portions of rat hepatocytes in mitosis and incubated for AcPase activity.

Upper figure: From a rat given nafenopin by gavage. Reaction product is seen in tubules and anastomosing cisternae of GERL (*arrows*) and lysosomes (*L*), which are positioned at the periphery of the mitotic apparatus. The Golgi apparatus is not evident. The Golgi apparatus virtually disappears or is reduced to a single cisterna (Novikoff et al., 1985a). Chromosomes are labeled *C*; peroxisomes, *P*; mitochondria, *M*. ×11,000.

Lower figure: From a ratfed swainsonine. Reaction product is seen only in the numerous lysosomes, which are closely apposed and possibly interconnected (*arrows*); they are clustered at the mitotic poles. The Golgi apparatus is not visible. Mitochondria, peroxisomes, and chromosomes are labeled *M*, *P*, and *C*, respectively. ×8000.

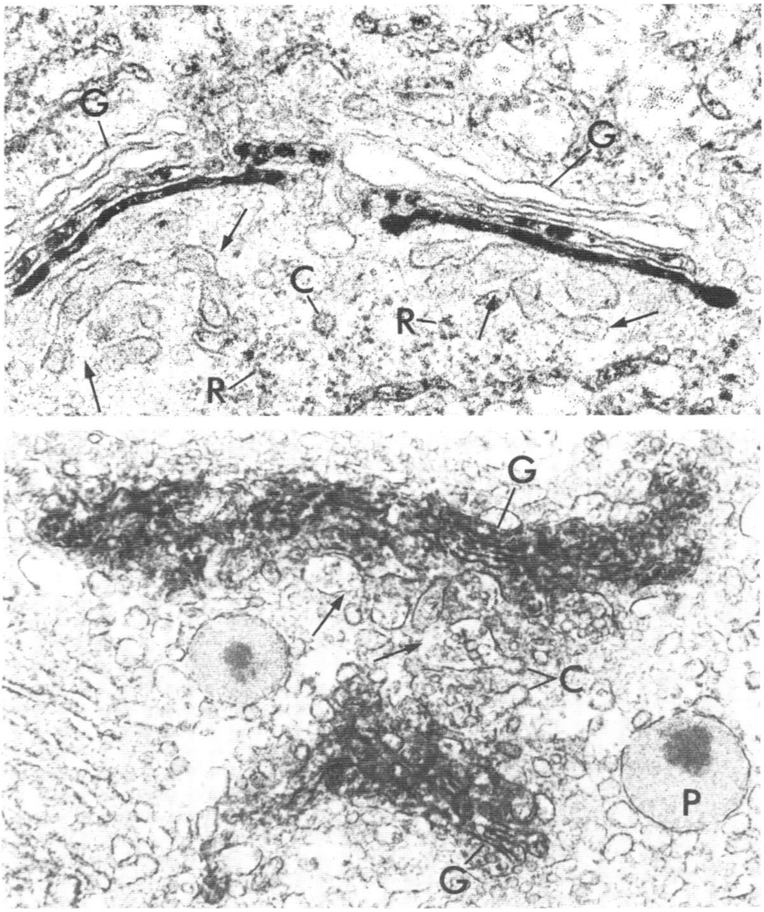

Figure 10. Portions of hepatocytes from untreated rat liver.

Upper figure: Incubated for TPPase activity. Only the trans elements of the Golgi stack (*G*) show activity. GERL (*arrow*) and coated vesicle (*C*) are without activity. Numerous free ribosomes (*R*) are seen in proximity to GERL. ×55,000.

Lower figure: Incubated for α-D-mannosidase activity. In contrast to TPPase, all the elements of the Golgi stack (*G*) show activity. GERL (*arrows*) is negative for this enzyme. Lipoprotein particles are evident in GERL, as are coated regions (*C*). ×40,000. (From Novikoff et al., 1983, with permission.)

This method stained only one or two cisternae in rat hepatocytes (Figure 10, upper). In contrast, α-D-mannosidase II is found in all the cisternae of the Golgi apparatus (Figure 10, lower). In light microscopic preparations of α-D-mannosidase localizations (Figure 7, upper), the extensive highly twisted and convoluted appearance of the entire apparatus is better appreciated. This micrograph also shows that this enzyme has a different distribution from acid mannosidase, which is localized in lysosomal structures (see Figures 3 and 7; Novikoff et al., 1985a).

V. BIOGENESIS OF LYSOSOMAL ENZYMES

The initial steps in the biosynthesis of lysosomal enzymes involves synthesis on membrane-bound polysomes and translocation into the endoplasmic reticulum (Walter et al., 1984; Palade, 1975; Sabatini et al., 1982; Blobel, 1982; Davis and Tai, 1980). After a number of modifications (e.g., glycosylation, phosphorylation, sulfation, proteolytic cleavage), the lysosomal enzymes are segregated into lysosomes by the interaction between mannose-6-phosphate membrane receptors and mannose-6-phosphate present on the newly synthesized acid hydrolases (Creek and Sly, 1984; Farquhar, 1985; Goldberg, 1984).

It is considered that secretory and lysosomal proteins follow similar biosynthetic pathways until the proteins are delivered into their respective organelles; that is, they are synthesized in the endoplasmic reticulum, then transported to and through the Golgi apparatus. After these steps, the pathway for secretory and lysosomal proteins diverge with secretory proteins incorporated into secretory vacuoles and lysosomal proteins into acid hydrolase-positive structures (lysosomes, GERL, and pleomorphic tubules).

The evidence for the transit of lysosomal enzymes through the Golgi apparatus rests mostly on the association of carbohydrate-processing enzymes in Golgi-enriched fractions isolated from liver (Fleischer et al., 1969; Morré et al., 1969; Schachter et al., 1970; Tabac and Kornfeld, 1979; Varki and Kornfeld, 1980; Tulsiani et al., 1982). It has been assumed that since most lysosomal enzymes are glycoproteins they are glycosylated in the Golgi apparatus in the same manner as secretory proteins. Low levels of acid hydrolase have been reported in Golgi-enriched fractions by biochemical analysis; however, this activity may result from contamination by disrupted lysosomes, microsomes, or other acid hydrolases containing membrane-delimited structures. Furthermore, glycosyltransferases and not acid hydrolases are employed as enzymatic markers to characterize the Golgi apparatus. Were the Golgi apparatus involved in concentrating and segregating lysosomal enzymes, one would expect to detect considerably more acid hydrolases in the Golgi-enriched fractions than is actually found.

Numerous *in situ* ultrastructural studies on the intracellular localizations of several lysosomal enzymes and on lysosomal membrane proteins have not found lysosomal enzymes in the Golgi apparatus (Bentfeld and Bainton, 1975; Decker, 1974; Novikoff et al., 1983, 1985a; Oliver, 1980; Decker et al., 1980; Iwamasa et al., 1982; Lewis et al., 1985). Moreover, light microscopic studies on the localization of AcPase in hepatocytes (and other cell types) display a dot-like lysosomal distribution rather than a ribbon-like Golgi apparatus distribution (Figure 7). Since a larger number of cells and a greater depth of cells can be seen in light microscopic preparations, the possibility of seeing acid hydrolases with a distribution similar to that of the Golgi apparatus would be enhanced.

Since lysosomal enzymes have not been detected in the Golgi apparatus, the lysosomal biosynthetic pathway may differ from the secretory pathway. If lysosomal enzymes bypass the Golgi apparatus, then their modifications after synthesis in the endoplasmic reticulum must occur at a different intracellular site from secretory proteins. The additional modifications may occur in specific regions of the endoplasmic reticulum, in GERL, or in other unidentified compartments. We have localized the Golgi enzyme, α-D-mannosidase, in regions of smooth and rough ER (Novikoff et al., 1983). It is known that some glycosylation of proteins occurs in the endoplasmic reticulum (Atkinson & Kabcenell, 1984). Recently sialyetransferase was localized to an AcPase-positive cisterna (GERL) (Roth et al., 1985). The notion that lysosomal enzymes do not transit the Golgi apparatus before their incorporation into lysosomes and other acid hydrolase membrane compartments was based mostly on the localization of one acid hydrolase, namely, AcPase.

A number of additional acid hydrolases have been localized by our laboratory and others, thereby providing further evidence that alternate lysosomal biosynthetic pathways may exist. Enzyme cytochemical methods available for the localization of lysosomal enzymes (particularly at the electron microscope level) are few in number (e.g., AcPase, aryl sulfatase, E-600-resistant esterase, trimetaphosphatase). With the development of valid and sensitive immunocytochemical methods and the production of highly specific antibodies, the number of acid hydrolases that could be demonstrated *in situ* was greatly expanded. We have localized β-galactosidase, β-glucuronidase, and acid mannosidase in lysosomes (residual bodies, autophagic vacuoles), GERL, smooth-surface tubules, and in rough and smooth endoplasmic reticulum. None of these lysosomal enzymes was detected in the Golgi apparatus. This lack of reactivity is not due to nonpenetrability of antibodies or other immunoreagents because we were able to immunolocalize the Golgi enzyme, α-D-mannosidase, in all the elements of the Golgi apparatus. Lysosomal enzymes were not found in any cisterna of the Golgi apparatus, even under conditions of optimal

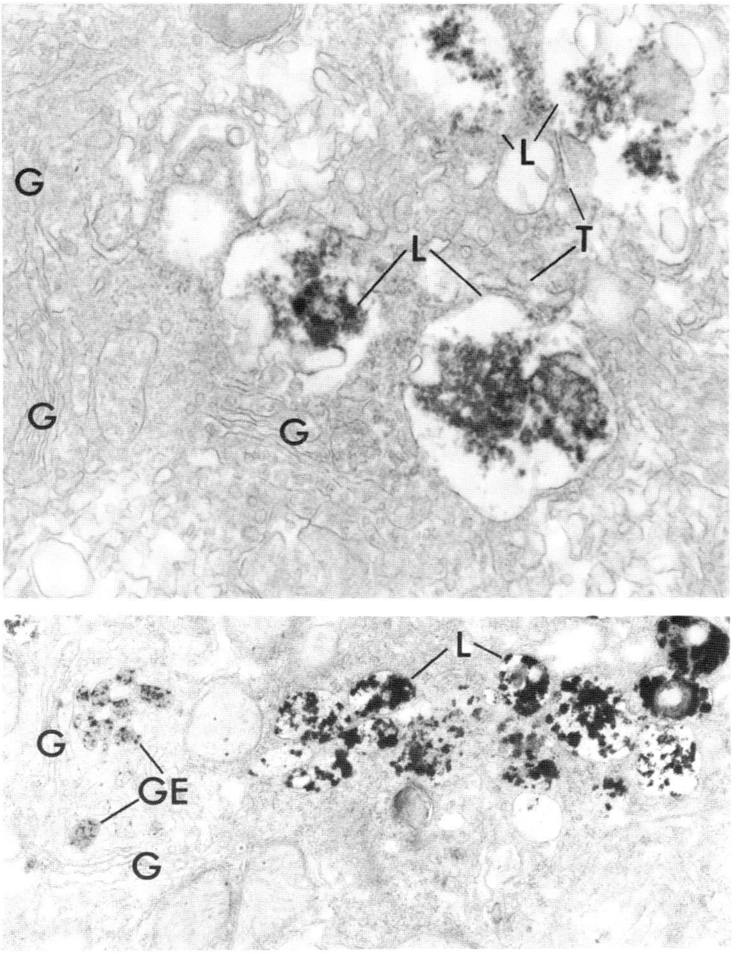

Figure 11. Portions of rat hepatocytes from livers of rats administered swainsonine.

Upper figure: Immunolocalization of acid mannosidase. Acid mannosidase activity is found in enlarged lysosomes (*L*) and in tubules (*T*). No activity is seen in the Golgi stack (*G*). ×40,000.

Lower figure: Localization of AcPase. This enzyme is localized to GERL (*GE*) and lysosomes, which are increased in number and size and are in close apposition, possibly connected to each other. The Golgi stack (*G*) is without AcPase activity. ×25,000. (From Novikoff et al., 1985a, with permission.)

antibody access (brief fixation; treatment with detergents; increased exposure time to antibody). Other investigators have not detected the aforementioned lysosomal enzymes, including cathepsin D and α-glucosidase, in the Golgi apparatus. Lysosomal enzyme activity was not observed in the Golgi apparatus even when secretory lipoprotein particles were seen in the Golgi cisternae. The absence of lysosomal enzymes in the Golgi apparatus cannot be due to nonfunctioning of the Golgi in transport of proteins; the presence of lipoprotein particles in the Golgi cisternae indicate that the Golgi apparatus was actively engaged in transport of secretory material. It is possible that the enzyme and immunocytochemical procedures are not sensitive enough to detect these enzymes. It is also possible that the lysosomal enzymes transit the Golgi so rapidly that they elude detection. However, it is unlikely that this could be the situation for every enzyme localized and for all the different cell types in which localizations have been performed. It is also highly unlikely that the transit of all lysosomal enzymes through the Golgi stack proceeds at the same rate or is so precisely synchronized or coordinated that none of the lysosomal enzymes are detected as they traverse the Golgi stack. Lysosomes contain high concentrations of lysosomal enzymes. The absence of these enzymes in the Golgi apparatus suggests that it is not involved in the transport, concentration, or packaging of lysosomal enzymes into lysosomes. If the Golgi apparatus is involved in these processes, one would expect to find at least gradations of lysosomal enzymes, with the highest concentration in the *trans*-Golgi cisternae, the lowest in the *cis*-Golgi cisternae. This has not been found.

We have also found the Golgi apparatus to be negative for lysosomal enzymes (AcPase and acid mannosidase) even under conditions when acid hydrolases are elevated and the number of lysosomes increases (Figure 11). This occurs in hepatocytes of livers from rats administered the plant alkaloid swainsonine. Swainsonine not only increases the lysosomal enzyme levels of acid mannosidase (Tulsiani and Touster, 1983) but also induces a hypertrophy of the lysosomal system (GERL and residual bodies) without causing a concomitant change in the morphological appearance of the Golgi apparatus (Novikoff et al., 1985a). The effects of swainsonine provide further evidence that the lysosomal system is a distinct structure from the Golgi apparatus and can be modulated without affecting the Golgi apparatus.

VI. THE LYSOSOMAL SYSTEM (ACID HYDROLASE COMPARTMENTS)

The existence of an extensive lysosomal compartment became evident to us when we localized AcPase and exogenously administered horseradish peroxidase in Kupffer cells and spleen cells (Novikoff et al., 1981). In these macrophages, the system consisted of interconnected spherical lysosomes (residual bodies, etc.) and elongated tubular lysosomes as well as cisternae of GERL. The localization of AcPase in these structures showed the lysosomal and continuous nature of the system. The presence of peroxidase showed that endocytosed macromolecules entered the system and revealed the extent and form of the system. The elongated tubular lysosomes also showed other characteristics of conventional lysosomes, namely, a clear region between the delimiting membrane and electron-dense matrix (see Figure 2B in Novikoff et al., 1981). In Kupffer cells and spleen cells lipoprotein particles, presumably of exogenous origin, were found within the system.

The lysosomal system is also found in hepatocytes. The initial observations of interconnected lysosomes occurred as early as 1966 (de Duve and Wattiaux, 1966); it was was not until 1973 (Novikoff, 1973) and 1978 (Novikoff and Shin, 1978) that micrographs were published showing autophagic vacuoles and residual bodies connected by tubules. Serial sections of rat hepatocytes revealed that lysosomes in the centrosphere area (Golgi zone) were not separate entities but were interconnected. In 1982, at a Gordon Conference on Lysosomes,[4] evidence was presented that an extensive lysosomal system is present in normal hepatocytes; electron micrographs were shown of acid hydrolase-rich structures (GERL, lysosomes, elongated tubules) localized not only in the Golgi zone but also in other parts of the hepatocyte cytoplasm. At that time it was suggested that acid hydrolase-rich compartments are not limited to the Golgi zone, but extend to other regions of the hepatocyte cytoplasm.

The upper panels in Figure 12 show AcPase-positive pleomorphic structures located in various regions of the hepatocyte. Elongated, twisted, and branching tubules are found in sinusoidal, lateral, nuclear, and Golgi regions of the cytoplasm. Continuities between tubules and spherical lysosomes are frequently encountered (Figure 12, lower). In hepatocytes of the mutant beige mouse an even more extensive lysosomal system is found. The upper panel and lower right panel in Figure 13 show the extensive tubular network in the sinusoidal region and the hypertrophy of GERL in the Golgi zone of hepatocytes from the beige mouse. Exogenous peroxidase is transported into the lysosomal system of rat and beige mouse hepatocytes. Figure 13, lower left, shows peroxidase, using diaminobenzidine cytochemistry (Graham and Karnovsky, 1966), in structures that are identical to those with AcPase activity. Previously, we reported that endocytosed receptor-mediated macromolecules (e.g., asialoglycoproteins labeled with electron-dense tracers) entered lysosomal compartments (Stockert et al., 1980; Haimes et al., 1981).

Unusual tubular lysosomes have been reported in parotid gland cells (Oliver, 1980), in glucagon-treated hepatocytes (Saita and Ogawa, 1974),

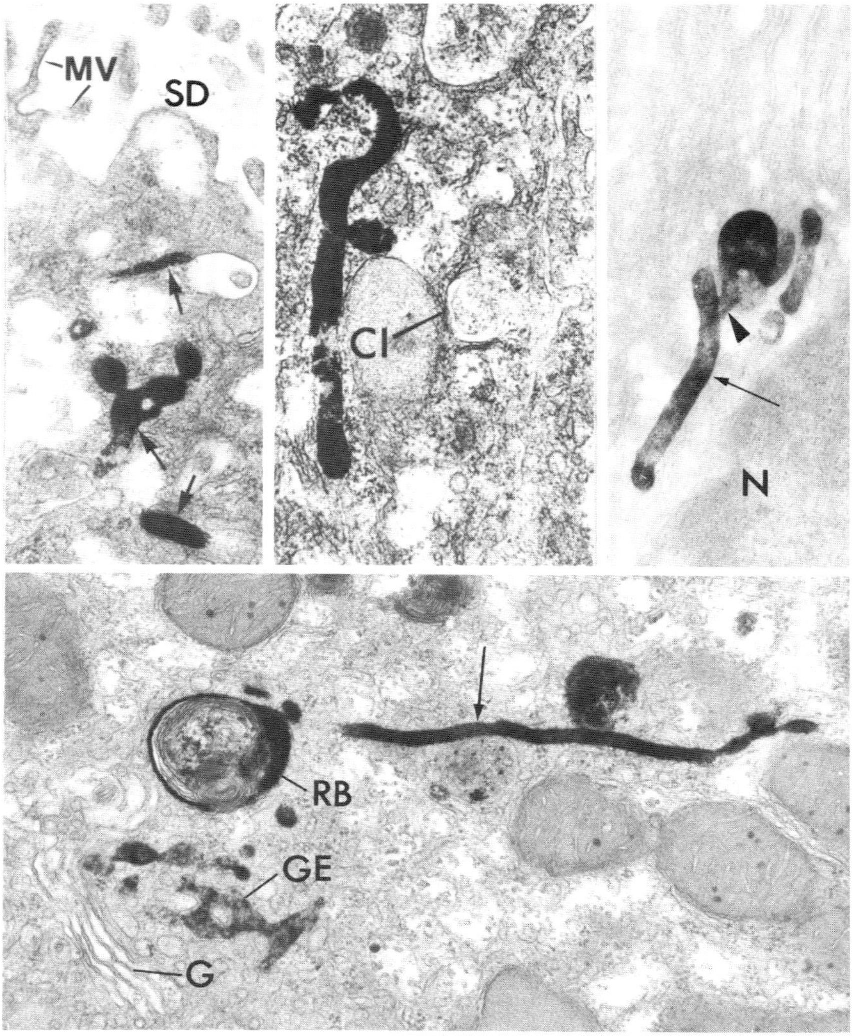

Figure 12. Portions of hepatocytes from livers of untreated rats incubated for AcPase activity using cytidine monophosphate as substrate. These figures illustrate pleomorphic cisternae and tubules of the acid hydrolase-rich compartments in different regions of the hepatocyte.

Upper figure: In the *left panel*, reaction product is seen in tubules (*arrows*) near the sinusoidal aspect of the hepatocyte. Microvilli (*MV*) and space of Disse (*SD*) are labeled. In the *center panel*, an elongated tubule is seen near the lateral plasma membrane (*C*1). In the *right panel*, tubules are seen near the nucleus (*N*). A connection between a tubule and a lysosome is seen at the arrowhead. *Left*, ×30,000; *center*, ×30,000; *right*, ×33,000.

in exocrine pancreas cells (Beaudoin et al., 1985), in peritoneal macrophages (Phaire-Washington et al., 1980), and in aortic smooth muscle cells (Robinson et al., 1986). These structures not only have the morphological characteristics of traditional spherical lysosomes (a tripartite delimiting membrane and dense matrix with a clear zone in between), but also were reactive for acid hydrolases. It was shown that the tubular lysosomes could also be demonstrated in living aortic smooth muscle cells, indicating that the elongated lysosomes were not artifacts of fixation (Robinson et al., 1986).

As a working hypothesis, we consider the lysosomal system to include smooth membrane delimited acid hydrolase structures present throughout the cytoplasm. This system could include spherical and elongated lysosomes (residual bodies, dense bodies, autophagic vacuoles, coated vesicles) and anastomosing cisternae and tubules of GERL located in the Golgi zone and pleomorphic cisternae and tubules, and lysosomes found elsewhere in the cytoplasm. Our impression is that acid hydrolase-rich compartments traverse all regions of the cytoplasm and occupy a larger volume of the cytoplasm than heretofore appreciated. The extent and continuity of this compartment has not been determined. High-voltage electron microscopic studies of thick sections incubated for acid hydrolases are in progress and are expected to reveal the three-dimensional appearance of this system (Novikoff et al., 1985). We do not know if the individual components of the system are permanently or temporarily connected to each other, if only specific regions are connected, or how these connections are influenced by altering physiological conditions.

Particularly, we would like to know the structural and functional relations of the lysosomal compartments to structures that are involved in fluid phase and receptor-mediated endocytosis (phagocytic vacuoles, endosomes, CURL,[5] receptosomes) (Straus, 1964; Steinman et al., 1983; Willingham et al., 1983; Helenius et al., 1983; Geuze et al., 1983). To determine the relation of these structures to the lysosomal system, we are presently conducting studies in hepatocytes that involve the simultaneous localization of electron-dense tracers and enzyme markers. Acid hydrolases will be used to delineate lysosomal structures, and electron-dense tracers, unlinked or linked to specific macromolecules, will be used to identify the endocytotic pathway. These double-labeled experiments will also determine the relation of GERL to the reticular elements at the *trans* side of the Golgi apparatus.

Figure 12—continued

Lower figure: An elongated tubule with a lysosome attached is seen in the Golgi zone. An anastomosing portion of GERL (*arrows*) is evident at the trans aspect of the Golgi stack (*G*). The tubules that course through the cytoplasm, GERL (*GE*), and lysosomes probably form an interconnected acid hydrolase-rich compartment comprising the lysosomal system. ×27,000.

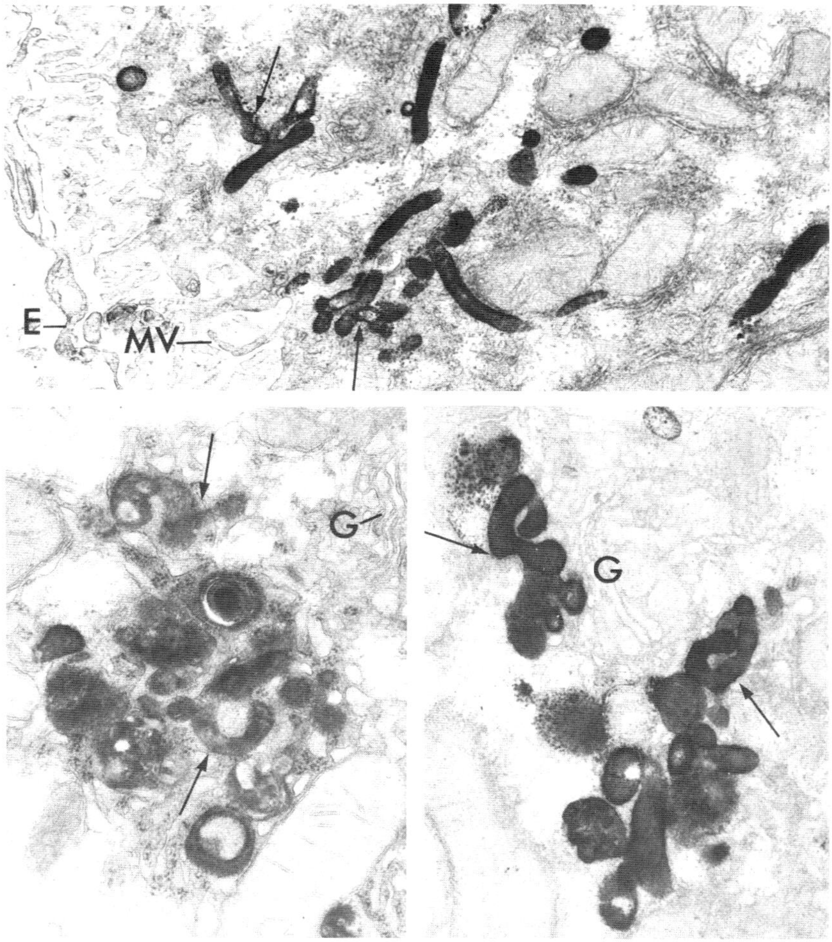

Figure 13. Portions of hepatocytes from the livers of beige mutant mouse injected with horseradish peroxidase for 20 minutes.

Upper figure: Incubated for trimetaphosphatase activity. Reaction product delineates numerous elongated tubules, some of which branch. Some are located in a region of the hepatocyte that is close to the sinusoidal surface. Microvilli (*MV*) are seen in the space of Disse. An endothelial cell (*E*) is also labeled. ×21,000.

Lower figure, left: Incubated in diaminobenzidine, pH 7.6. GERL (*arrows*) and lysosomes show peroxidase activity. Note the absence of peroxidase activity in the Golgi apparatus (*G*). ×25,000.

It is also interesting to speculate on the significance of the additional AcPase compartments in other parts of the cytoplasm in relation to our present knowledge on the sorting of lysosomal enzymes by mannose-6-phosphate receptors. These receptors have been shown to be localized in either specific cisternae of the Golgi apparatus or in cisternae of GERL (so far, there is no agreement on the precise localization of these receptors [see Farquhar, 1985, for discussion of this issue]). However, since lysosomal enzymes are now known to be present in other areas of the cytoplasm, there may be additional sorting sites by this or other receptors as well as other intracellular biosynthetic pathways for lysosomal enzymes. It would also be of interest to determine whether the acid hydrolase structures outside the centrosphere region of the cell represent sites of degradative events (e.g., autophagy).

We have demonstrated that the ER is continuous with acid hydrolase structures in the Golgi zone and that numerous free ribosomes are close to these structures, suggesting that synthetic events may be associated with these structures. Endoplasmic reticulum has also been observed in proximity to other components of the lysosomal system.

VII. CONCLUSIONS

We have witnessed a dramatic burst of knowledge about acid hydrolase compartments since the lysosome and GERL concepts were proposed. Lysosomes were often viewed as separate structures, and only since evidence that lysosomes are connected to each other have investigators been thinking in terms of a lysosomal system. The extensive nature of this system is beginning to be appreciated with the demonstration that acid hydrolase compartments are not confined to the centrosphere or Golgi region of the cell but extend into all regions of the cell's cytoplasm via numerous linear or pleomorphic tubular structures. The presence of an extensive lysosomal system has implications not only for degradative events but also for biosynthesis, transport, and sorting of lysosomal enzymes. The relation of this system to structures involved in endocytotic and other transcellular pathways

---

Figure 13—continued

Lower figure, right: Incubated for trimetaphosphatase (TMPase) activity. Reaction product is found in GERL (*arrows*) and lysosomes. The Golgi apparatus does not display any TMPase activity. The structures with TMPase activity and those with peroxidase activity are similar, indicating that exogenous tracers enter the acid hydrolase-rich compartments. ×20,000.

is presently under investigation. Of particular interest is the function of the components or subcompartments of the lysosomal system in synthetic events of the cell; this will soon be possible with *in situ* cDNA hybridization cytochemistry. We look forward with much excitement to uncovering the multiple functions of the acid hydrolase compartments of the cell.

ACKNOWLEDGMENTS

I gratefully acknowledge the collaboration of Alex Novikoff, without whom much of this research could not have been done. I am also thankful to my collaborators at Albert Einstein College of Medicine and at other institutions. Much appreciation is also extended to Mr. Harry Madara, who prepared the photomicrographs, and to Mrs. Fay Grad, who typed the final version of the manuscript. This research was supported by National Institutes of Health Grants AM 23078 and CA 06576.

NOTES

1. O. L. Thomas, Quart. J. Microscop. Sci. *104:* 75-79 (1963).
2. A. B. Novikoff, and S. Goldfischer, Proc. Natl. Acad. Sci. USA *47:* 802-810 (1961).
3. A. B. Novikoff, Jewish Mem. Hosp. Bull. *6:* 70-93 (1962).
4. P. M. Novikoff, slide presentation at the Gordon Conference on Lysosomes, Andover, NH, 1982.
5. CURL, compartment for uncoupling receptor and ligand.

REFERENCES

Alexander, C. A., Hamilton, R. L., and Havel, R. J. (1976) J. Cell Biol. *69:* 241-263.

Atkinson, P. H., and Kabcenell, A. K. (1984) Biochem. Soc. Trans. *12:* 519-521.

Baker, J. R. (1963) J. R. Microsc. Soc. *82:* 145-157.

Beaudoin, A. R., Grondi, G., Lord, A., and Pelletier, M. (1985). J. Histochem. Cytochem. *33:* 569-575.

Bennett, G., and O'Shaughnessy, D. (1981) J. Cell Biol. *88:* 1-15.

Bentfeld, M. E., and Bainton, D. F. (1975) J. Clin. Invest. *56:* 1635-1649.

Blobel, G. (1982) *The Harvey Lectures* Ser. 76, pp. 125-147. Academic Press, New York.

Boutry, J.-M., and Novikoff, A. B. (1975) Proc. Natl. Acad. Sci. USA *72:* 508-512.

Cheetham, D., Morré, J., Pannek, C., and Friend, D. S. (1971) J. Cell Biol. *49:* 899-905.

Creek, K. E., and Sly, W. S. (1984) *Lysosomes in Biology and Pathology* (J. T. Dingle, R. T. Dean, and W. Sly, eds.), Vol. 7, pp. 63-82. Elsevier/North-Holland, New York.

Dalton, A. J., and Felix, M. D. (1956) J. Biophys. Biochem. Cytol. *2,* (Suppl.): 79-83.

Davis, B. D., and Tai, P.-C. (1980) Nature (Lond.) *283:* 433-438.

de Duve, C. (1975) The Nobel Foundation (Nobel Lecture), pp. 139-160.

de Duve, C. (1969) in *Lysosomes in Biology and Pathology* (J. T. Dingle and H. B. Fell, eds.) Vol. 1, pp. 3-40. Elsevier/North-Holland, New York.

de Duve, C., Pressman, B. C., Gianetta, R., Wattiaux, R., and Appelman, F. (1955) Biochem. J. *60:* 604–617.
de Duve, C., and Wattiaux, R. (1966) Annu. Rev. Physiol. *28:* 435–492.
Decker, R. S. (1974) J. Cell Biol. *61:* 599–612.
Decker, R. S., Decker, M. L., and Poole, A. R. (1980) J. Histochem. Cytochem. *28:* 231–237.
Dunn, W. A., Connolly, T. P., and Hubbard, A. L. (1986) J. Cell Biol. *102:* 24–36.
Dunphy, W. G., Brands, R., and Rothman, J. E. (1985) Cell *40:* 463–472.
Eder, H. A., Novikoff, P. M., Novikoff, A. B., Yam, A., Beyer, M., and Gidez, L. I. (1979) Proc. Natl. Acad. Sci. USA *76:* 5905–5909.
Erikson, A. H., Conner, G. F., and Blobel, G. (1981) J. Biol. Chem. *256:* 11224–11231.
Essner, E., and Haimes, H. (1979) J. Cell Biol. *75:* 381–387.
Essner, E., and Oliver, C. (1974) Lab. Invest. *30:* 596–607.
Essner, E., and Novikoff, A. B. (1961) J. Biophys. Biochem. Cytol. *9:* 773–784.
Farquhar, M. G. (1985) Annu. Rev. Cell Biol. *1:* 447–488.
Farquhar, M. G. (1978) J. Cell Biol. *77:* R35–R42.
Farquhar, M. G. (1969) in *Lysosomes in Biology and Pathology* (J. T. Dingle and H. B. Fell, eds.), Vol. 2, pp. 463–482. Elsevier/North-Holland, New York.
Fleischer, D., Fleischer, S., and Ozawa, H. (1969) J. Cell Biol. *43:* 59–79.
Garvin, A. J., Lyubsky, S., and Poore, C. M. (1981) Exp. Cell Res. *133:* 297–307.
Geuze, H. J., Slot, J. W., Strous, G. J., Lodish, H. F., and Swartz, A. L. (1983) Cell *32:* 277–287.
Gieselmann, V., Pohlmann, R., Hasilik, A., and Von Figura, K. (1983) J. Cell Biol. *97:* 1–5.
Golgi, C. (1898) Arch. Ital. Biol. *30:* 278–286.
Goldberg, D., Gabel, G., and Kornfeld, S. (1984) in *Lysosomes in Biology and Pathology* (J. T. Dingle, R. T. Dean, and W. Sly, eds.), Vol. 7, pp. 45–62. Elsevier/North-Holland, New York.
Gonatas, N. K., Kim, S. U., Stieber, A., and Avrameas, S. (1977) J. Cell Biol. *73:* 1–13.
Graham, R. C., Jr., and Karnovsky, M. J. (1966) J. Histochem. Cytochem. *14:* 291–302.
Griffiths, G., Pfeiffer, S., Simons, K., and Matlin, K. (1985) J. Cell Biol. *101:* 949–964.
Hand, A. R., and Oliver, C. (1977a) J. Cell Biol. *74:* 399–413.
Hand, A. R., and Oliver, C. (1977b) Histochem. J. *9:* 375–392.
Haimes, H., Stockert, R. J., Morell, A. G., and Novikoff, A. B. (1981) Proc. Natl. Acad. Sci. USA *78:* 6936–6939.
Helenius, A., Mellman, I., Wall, D., and Hubbard, A. (1983) Trends Biochem. Sci. *8:* 245–250.
Holt, S. J., and Hicks, R. M. (1961) J. Biophys. Biochem. Cytol. *11:* 47–66.
Holtzman, E., and Dominitz, R. (1968) J. Histochem. Cytochem. *16:* 321–336.
Holtzman, E., Novikoff, A. B., and Villaverde, H. (1967) J. Cell Biol. *33:* 419–436.
Iwamasa, T., Nobuyuki, N., Hmada, T., and Fukuda, S. (1982) J. Histochem. Cytochem. *30:* 378–384.
Jaeken, L., Thines-Sempoux, D., and Verheyen, F. (1978) Cell Biol. Int. Rep. *2:* 501–513.
Joseph, K. C., Kim, S. U., Stieber, A., and Gonatas, N. K. (1978) Proc. Natl. Acad. Sci. USA *75:* 2815–2819.
Kuhn, N. J., and White, A. (1977). Biochem. J. *168:* 423–433.
Lacy, D., and Challice, C. E. (1957) Soc. Exp. Biol. Symp. *10:* 62–91.
Lewis, V., Green, S. A., Marsh, M., Vikho, P., Helenius, A., and Mellman, I. (1985) J. Cell Biol. *100:* 1839–1847.
Marty, F. (1978) Proc. Natl. Acad. Sci. USA *75:* 381–387.
Mollenhauer, H. H., and Morré, D. J. (1966) Annu. Rev. Plant Physiol. *17:* 27–46.
Mori, M., Satoh, M., and Novikoff, A. B. (1982) Exp. Cell Res. *141:* 277–282.
Morré, D. J., Merlin, L. M., Keenan, T. W. (1969) Biochem. Biophys. Res. Commun. *37:* 813–819.
Nehemiah, J. L., and Novikoff, A. B. (1974) Exp. Mol. Pathol. *21:* 398–423.
Novikoff, A. B. (1985) Curr. Contents, Citation Classic *28:* 19.

Novikoff, P. M. (1977) Proc. Natl. Acad. Sci. USA *74:* 3550–3557.
Novikoff, A. B. (1973) in *Lysosomes and Storage Diseases* (G. Hers and F. Van Hoof, eds.), pp. 1–41. Academic Press, New York.
Novikoff, A. B. (1964) Biol. Bull. *127:* 358.
Novikoff, A. B., and Goldfischer, S. (1961) Proc. Natl. Acad. Sci. USA *47:* 802–810.
Novikoff, A. B., and Novikoff, P. M. (1977) Histochem. J. *9:* 525–551.
Novikoff, P. M., and Novikoff, A. B. (1972) J. Cell Biol. *53:* 532–560.
Novikoff, A. B., and Shin, W.-Y. (1978) Proc. Natl. Acad. Sci. USA *75:* 5039–5042.
Novikoff, P. M., and Yam, A. (1978a) J. Cell Biol. *76:* 1–11.
Novikoff, P. M., and Yam, A. (1978b) J. Histochem. Cytochem. *26:* 1–13.
Novikoff, A. B., Beaufay, H., and de Duve, C. (1956) J. Biophys. Biochem. Cytol. *2:* 65–66.
Novikoff, P. M., and Edelstein, D. (1977) Lab Invest. *36:* 215–231.
Novikoff, P. M., LaRusso, N. F., Novikoff, A. B., Stockert, R. J., Yam, A., and Le Sage, G. D. (1983) J. Cell Biol. *97:* 1559–1565.
Novikoff, A. B., Leuenberger, P. M., Novikoff, P. M., and Quintana, N. (1979) Lab. Invest. *40:* 155–165.
Novikoff, A. B., Novikoff, P. M., Mori, M., and Levine, W. (1975) J. Histochem. Cytochem. *23:* 314–315.
Novikoff, A. B., Mori, M., Quintana, N., and Yam, A. (1977) J. Cell Biol. *75:* 148–165.
Novikoff, P. M., Novikoff, A. B., and Ris, H. (1985) J. Cell. Biol. *101:* (Pt. 2): 55a.
Novikoff, P. M., Roheim, P. S., Novikoff, A. B., and Edelstein, D. Lab Invest. *30:* 732–750.
Novikoff, A. B., Novikoff, P. M., Rosen, O. M., and Rubin, C. S. (1980) J. Cell Biol. *87:* 180–196.
Novikoff, P. M., Novikoff, A. B., Quintana, N., and Davis, C. (1973a) J. Histochem. Cytochem. *21:* 540–558.
Novikoff, A. B., Novikoff, P. M., Quintana, N., and Davis, C. (1973b) J. Histochem. Cytochem. *21:* 1010–1020.
Novikoff, P. M., Novikoff, A. B., Quintana, N., and Hauw, J.-J. (1971) J. Cell Biol. *50:* 859–886.
Novikoff, P. M., Touster, O., Novikoff, A. B., and Tulsiani, D. R. P. (1985a) J. Cell Biol. *101:* 339–349.
Novikoff, P. M., Tulsiani, D. R. P., Touster, O., Yam, A., and Novikoff, A. B. (1983) Proc. Natl. Acad. Sci. USA *80:* 5364–5368.
Novikoff, P. M., Yam, A., and Novikoff, A. B. (1981) Proc. Natl. Acad. Sci. USA *78:* 5699–5703.
Novikoff, A. B., Yam, A., and Novikoff, P. M. (1975) Proc. Natl. Acad. Sci. USA *72:* 4501–4505.
Oliver, C. (1980) J. Histochem. Cytochem. *28:* 78–81.
Oliver, C., and Essner, E. (1973) J. Histochem. Cytochem. *21:* 218–228.
Palade, C. (1975) Science (Wash. DC) 347–358.
Palay, S. L. (1956) in *Progress in Neurobiology*, Vol. I: *Neurochemistry* (S. Korey and J. I. Nurnberger, eds.) pp. 64–82. Harper & Row, New York.
Palay, S. L., and Palade, G. E. (1955) J. Biophys. Biochem. Cytol. *1:* 69–88.
Pelletier, C., and Novikoff, A. B. (1972) J. Histochem. Cytochem. *20:* 1–12.
Phaire-Washington, L., Siverstein, S. C., and Wang, E. (1980) J. Cell Biol. *86:* 641–655.
Pino, R. M., Pino, L. C., and Bankston, P. W. (1981) J. Histochem. Cytochem. *29:* 1061–1070.
Reitman, M. L., and Kornfeld, S. (1981) J. Biol. Chem. *256:* 4275–4281.
Renston, R. H., Maloney, D. G., Jones, A. L., Hradek, G. T., Wong, K. Y., and Goldfine, I. D. (1980) Gastroenterology *78:* 1373–1388.
Robbins, E., Marcus, P. I., and Gonatas, N. K. (1964) J. Cell Biol. *21:* 49–62.
Rome, L. H., Garvin, A. J., Allietta, M. M., and Neufeld, E. F. (1979) Cell *17:* 143–153.
Robinson, J. M., Okada, T., Castellot, J. J., Jr., and Karnovsky, M. J. (1986) J. Histochem. Cytochem. in press.
Rosenfield, M. G., Kreibich, G., Popov, D., Kato, K., and Sabatini, D. D. (1982) J. Cell Biol. *93:* 135–143.
Roth, J., and Berger, E. G. (1982) J. Cell Biol. *93:* 223–229.

Roth, J., Taatjes, D. J., Lucocq, J. M., Weinstein, J., and Poulson, J. C. (1985) Cell *43:* 287–295.
Rouiller, C. (1960) Int. Rev. Cytol. *9:* 227–272.
Sabatini, D. D., Kreibich, G., Morimato, T., and Adesnik, M. (1982) J. Cell Biol. *92:* 1–22.
Saita, T., and Ogawa, K. (1974) Acta Histochem. Cytochem. *7:* 1–18.
Schachter, H., Jabbal, I., Hudgin, R. L., and Pinteric, L. (1970) J. Biol. Chem. *245:* 1090–1100.
Smith, C. E. (1980) J. Histochem. Cytochem. *28:* 16–26.
Steinman, R. M., Mellman, I. S., Muller, W. A., and Cohn, Z. A. (1983) J. Cell Biol. *98:* 29–36.
Stockert, R. J., Haimes, H. B., Morell, A. B., Novikoff, P. M., Novikoff, A. B., Quintana, N., and Sternlieb, I. (1980) Lab. Invest. *43:* 556–563.
Strous, W. (1964). J. Cell Biol. *20:* 497–502.
Tabac, I., and Kornfeld, S. (1979) J. Biol. Chem. *254:* 11655–11663.
Tulsiani, D. R. P., Harris, T. M., and Touster, O. (1982) J. Biol. Chem. *257:* 7936–7939.
Tulsiani, D. R. P., and Touster, O. (1983) Arch. Biochem. Biophys. *224:* 594–600.
Varki, A., and Kornfeld, S. (1980) J. Biol. Chem. *255:* 8398–8401.
Wagner, R. R., and Cynkin, M. A. (1969) Biochem. Bioiphys. Res. Commun. *35:* 139–150.
Walter, P., Gilmore, R., and Blobel, G. (1984) Cell *38:* 5–8.
Willingham, M. C., Pastan, I. H., and Sahagian, G. G. (1983) J. Histochem. Cytochem. *31:* 1–11.
Wood, L., and Kaplan, A. (1985) J. Cell Biol. *101:* 2063–2069.

GLYCOPROTEIN DYNAMICS IN THE *CHLAMYDOMONAS* FLAGELLAR MEMBRANE

Robert A. Bloodgood

I. CILIA AND FLAGELLA SERVE MANY FUNCTIONS

The existence of eukaryotic cilia and flagella and their motile behavior were first observed by Antonie van Leeuwenhoek around 1676. Since that time, a great deal of attention has been paid to cilia and flagella as cellular organelles that either propel cells through their surrounding medium (in the case of ciliated protozoa or spermatozoa) or propel materials across the surface of a sheet of ciliated cells (in the case of water over mollusk gills, mucus along the mammalian respiratory tract, or ova along the mammalian oviduct). Although the organelles themselves have been classified as cilia or flagella and their pattern of bend propagation classified as ciliary or flagellar motility, it is now known that both forms of beating motility are based on the sliding of outer doublet microtubules relative to one another

Advances in Cell Biology, Vol. 1, pages 97–130.

ISBN: 0-89232-792-8

due to the cyclic activity of adenosinetriphosphatase(ATPase)-containing dynein arms that interconnect adjacent doublet microtubules (Gibbons, 1981). Indeed, changes in the free calcium concentration surrounding the axoneme can induce the same organelles to switch between a "ciliary" and a "flagellar" pattern of motility (Bessen et al., 1980; Hyams and Borisy, 1978). Since a preparation of *axonemes* (the nonionic detergent-insoluble fraction of cilia or flagella) can be reactivated and will express this motility *in vitro*, much attention has been directed toward the study of the axoneme and relatively little attention has been directed toward the surfaces of cilia and flagella.

Eukaryotic cilia and flagella are cellular organelles that are enclosed in an extension of the cell's plasma membrane. In certain ways, cilia and flagella can be considered as separate cellular compartments in that the soluble fraction of cilia and flagella is not identical with the general cell cytoplasm and the ciliary or flagellar plasma membrane is not identical to the rest of the cell's plasma membrane. The issue of localized barriers within the plasma membrane as a means of establishing domains is a topic in cell biology that is currently attracting a great deal of attention (Gumbier and Louvard, 1985); although little is currently known about the basis for membrane and cytoplasmic compartmentalization in cilia and flagella, these organelles promise to be useful systems in which to attack this important problem. There are unique structural specializations associated with the base of the flagellum: the basal body, the ciliary necklace and cross-bridges interconnecting these two structures (Ringo, 1967; Dentler, 1981; Melkonian, 1984). Ciliary and flagellar surfaces possess a number of obvious specializations; many of them possess some surface coat that can take the form of an unstructured glycocalyx, filamentous projections, or scales (Figure 1) (Manton, 1965; Dentler, 1981; Melkonian, 1984). The surfaces of many algal, fungal, and protozoan flagella possess long and sometimes very elaborate flagellar hairs referred to as *mastigonemes* (Bouck, 1972); in at least some cases, it appears that these structures serve to reverse the direction of effective thrust resulting from the propagation of bends along the flagellum (Jahn et al., 1964). Freeze-etch studies have revealed a wide variety of intramembrane particle arrays unique to flagellar membranes (Bardele, 1981; Dentler, 1981; Manton, 1965; Weiss et al., 1977).

Although not generally recognized, ciliary and flagellar surfaces exhibit a number of specialized functions, not all of which are related to the basic motility of these organelles (Dentler, 1981). Modified cilia on interstitial Turbellarian worms serve an adhesive function in anchoring the organism to surfaces (Tyler, 1973). The swollen distal tip of the flagellar membrane of *Crithidia* and *Trypanosoma* serve to anchor these parasitic protozoa to the gut wall of their insect hosts (Brooker, 1971; Molyneux, 1969; Vickerman, 1973). Whooping cough involves the binding of the bacterium

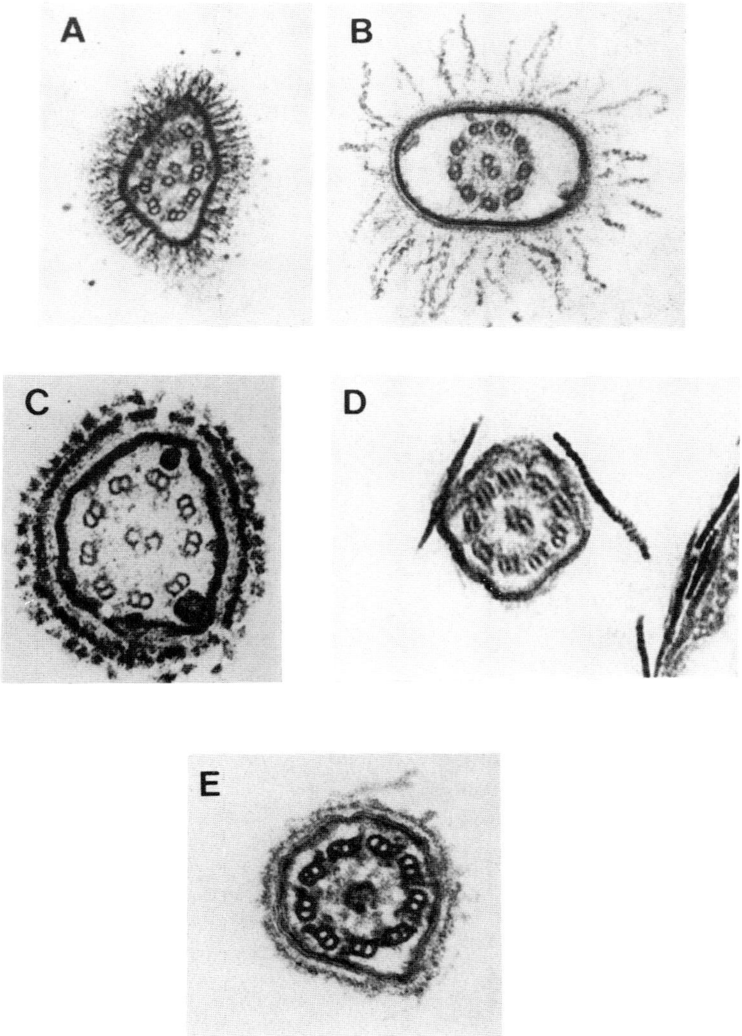

Figure 1. These electron micrographs display the enormous range of flagellar surface specializations observed among the algae: (A) *Chlorogonium*; (B) *Haematococcus*, (C) *Heteromastix*; (D) *Micromonas*; (E) *Chlamydomonas*. (Panels 1A, B, C, are reproduced from Manton, 1965; panel D is courtesy of Dr. Jeffrey Salisbury; panel E is reproduced from Bloodgood and May, 1982.) A, ×50,000; B, ×55,000; C, ×80,000; D, ×60,000; E, ×65,000.

Bordetella pertussis to specific regions of the membrane of respiratory tract cilia (Tuomanen and Hendley, 1983). The initial interactions of gametes during the mating of *Paramecium* and *Chlamydomonas* involve specific adhesion of ciliary and flagellar membranes, respectively (Adair, 1985; Kitamura and Hiwatashi, 1984; Snell, 1985; van den Ende, 1985; Watanabe, 1977). *Paramecium* ciliary membranes are involved in water transport (Ciccolello and Gibor, 1978), whereas mammalian olfactory ciliary membranes have been shown to possess binding sites for odorant molecules (Rhein and Cagan, 1980). Although modified cilia are an integral part of many sensory receptor systems (Barber, 1974; Moran et al., 1977), motile cilia can also function as sensory receptors. Ciliary and flagellar membranes possess voltage-dependent calcium channels (Dunlap, 1977; Machemer and Ogura, 1979) that can regulate the entry of external calcium into cilia and flagella. When swimming ciliate protozoa encounter an object, the deformation of the cilia results in a calcium action potential in the ciliary membrane (Kung and Saimi, 1982; Lee, 1978); the resultant increase in free-calcium concentration around the axoneme induces a modification in the bend pattern (ciliary reversal), and the organism then reverses its direction of swimming. Calcium fluxes across ciliary and flagellar membranes may be involved in other cellular processes such as the flagellar membrane signaling phenomenon occurring during *Chlamydomonas* mating (Bloodgood and Levin, 1983; Snell, 1985; Snell et al., 1982).

This incomplete list of functions performed by ciliary and flagellar surfaces demonstrates that these organelles are more versatile than is generally recognized and supports the argument that more emphasis should be placed on the study of ciliary and flagellar membranes. The *Chlamydomonas* flagellum is one system where considerable attention has been given to the study of the flagellar membrane in terms of adhesive, motile, and signaling events during the mating of gametes and in relation to the dynamics of vegetative *Chlamydomonas* flagellar membrane glycoproteins. The rest of this review will summarize the current state of our knowledge about this particular system in addition to providing some speculation in those areas where our information is still very incomplete.

II. ORGANIZATION OF THE *CHLAMYDOMONAS* FLAGELLAR SURFACE

Chlamydomonas is a eukaryotic, unicellular, green algal cell that possesses two flagella 10–12 μm in length. The cell body is covered by a multilayered cell wall that is made up, at least in part, of hydroxyproline-rich glycoproteins (Roberts et al., 1985). Although a great deal of work has been done using the flagella of *C. eugametos* in the laboratory of van den Ende (1985), most of the work to be described will apply to the flagella of *C. reinhardtii* unless otherwise indicated. The flagella of *C. reinhardtii* emerge from the cell body through two tunnels in the cell wall that are lined by flagellar collars (Snell, 1983). While continuous with the rest of the cell's plasma membrane, the flagellar membrane has properties that make it unique. The entire surface of the flagellar membrane (but not the cell body plasma membrane) is covered by a glycocalyx, or fuzzy coat, 20 nm in thickness (Figures 1E and 2A–C) (Ringo, 1967; Witman et al., 1972). Mastigonemes (0.9 μm long) composed of a single glycoprotein species of 170 kd (Witman et al., 1972; Snell, 1976) extend from the flagellar surface of *Chlamydomonas*; although they are seen along the entire length of the flagellum (Ringo, 1967), they may not be distributed around the entire circumference of the flagellum. The highly structured mastigonemes on *Euglena* (Bouck et al., 1978) and *Ochromonas* (Markey and Bouck, 1977) flagella remain mechanically attached to the axoneme after membrane removal. *Chlamydomonas* mastigonemes, although much simpler in organization than *Ochromonas* and *Euglena* mastigonemes, appear to be similarly anchored to the axoneme (Figure 2D). A number of intramembrane particle arrays have been reported to be associated with the flagellar membrane of *Chlamydomonas*. Longitudinal rows of large intramembrane particles extend the length of flagella in both vegetative and gametic cells (Bergman et al., 1975; Snell, 1976); similar longitudinal rows of intramembrane particles have been observed in other cilia and flagella (Bardele, 1981) and appear to be arranged so that one row overlies each outer doublet microtubule. Two other specialized arrays of intramembrane particles are located at the base of the flagellum. These are referred to as the flagellar necklace and the flagellar bracelet (Weiss et al., 1977) and could play a role in maintaining the flagellar membrane as a separate plasma membrane domain (Gumbiner and Louvard, 1985). The ciliary necklace is the site at which one observes prominent links between the flagellar plasma membrane and the basal body (Ringo, 1967). In addition, there are less obvious links between the flagellar membrane and the axoneme all along the length of the flagellum (Ringo, 1967; Bloodgood, 1977) (Figures (2A–C). Transmission electron micrographs of cross-sections of slightly swollen flagella taken by Dr. Greg May (Figure 2B) suggest that at least some of the slender membrane links may connect to the outer dynein arms. This type of configuration has been observed in one other system, the sensory cilia in insect and crustacean sensory receptors (Crouau, 1980, 1983). It is interesting that Marchese-Ragona and Johnson (1985) observed the localization of a monoclonal antibody made to *Tetrahymena* 14-S dynein to a location consistent with a ciliary membrane–axoneme link. Dentler et al. (1980) have published data suggesting that dynein-like ATPase is associated with *Tetrahymena* and mollusk gill ciliary membranes. These observations suggest that attention

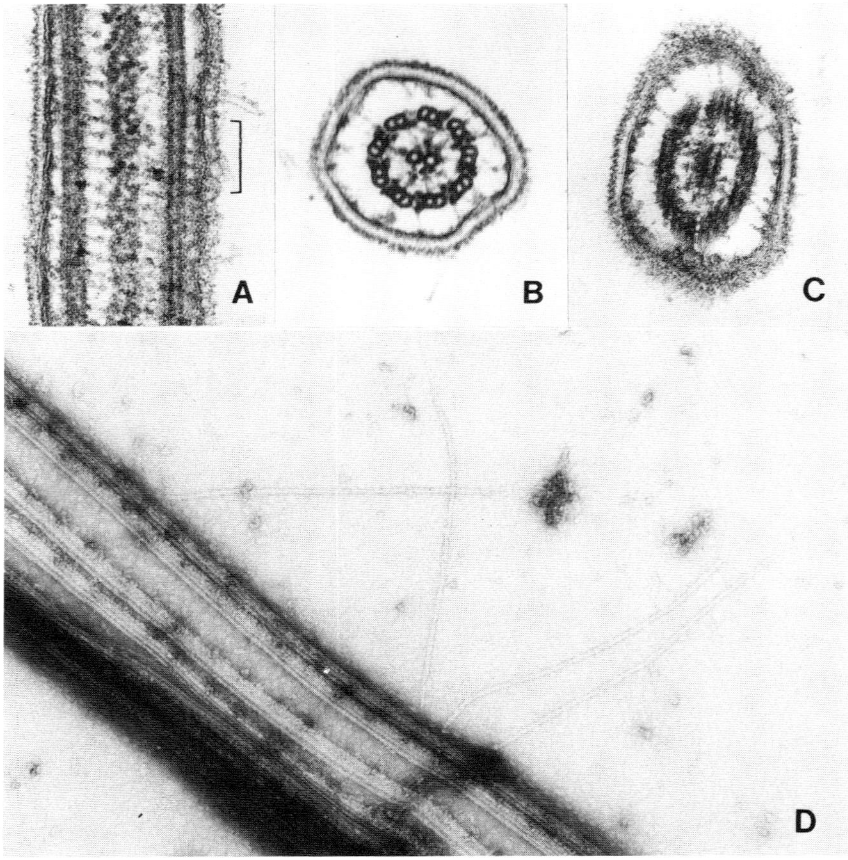

Figure 2. Electron micrographs of *Chlamydomonas reinhardtii* flagella.

(A) Longitudinal section through a flagellum from a strain of pf-18 cell showing the flagellar fuzzy coat and a series of membrane-axoneme links indicated by the bracket. Above the bracket are mastigonemes emanating from the flagellar surface. ×82,000.

(B) and (C) Cross sections of flagella from strain 21gr cells fixed in glutaraldehyde-containing tannic acid. The flagellar fuzzy coat is prominent; in addition, there are a number of links extending between the axoneme and the inside surface of the flagellar membrane. In panel B, it appears that some of the membrane–axoneme links may be contacting the outer dynein arms. Both, ×73,800.

(D) Negative stain image of an axoneme obtained by the extraction of *C. reinhardtii* strain 21gr flagella with the nonionic detergent Nonidet P-40. Note that several mastigonemes are still attached to the outer doublet microtubules in the total absence of the flagellar membrane. (Micrograph courtesy of Dr. L. I. Binder.) ×60,000.

should be paid to examining the possible role of dynein-ATPases in flagellar surface dynamic phenomenon. B. Daniel and C. King of University College, London (unpublished observations) found that mutant cell lines of *C. reinhardtii* lacking either the inner (pf-23) or the outer (pf-13a) dynein arms were deficient in gliding motility and exhibited a reduced area of flagellar contact with the substrate as judged by reflection interference microscopy. On the other hand, D. R. Mitchell and J. L. Rosenbaum (unpublished results) have looked for and found microsphere motility in a mutant of *C. reinhardtii* (pf28) totally lacking in outer dynein arms (Mitchell and Rosenbaum, 1985a).

Since flagella can be easily purified (Witman et al., 1972) and since there are no internal membranes within the flagella, it has been possible to examine the protein composition of *Chlamydomonas* flagellar membranes. The acrylamide gel pattern of isolated flagellar membranes or nonionic detergent extracts of whole flagella of *C. reinhardtii* is surprisingly simple and is dominated by a high-molecular-weight (HMW) glycoprotein region (made up of at least two components) with apparent molecular weight of 350 kd. These glycoproteins are present in both vegetative and gametic flagella (Bergman et al., 1975; Snell, 1976; Witman et al., 1972), are the major proteins exposed at the flagellar surface (Bloodgood and May, 1982; Bloodgood and Workman, 1984; Monk et al., 1983), represent the principal flagellar membrane components that interact with a substrate during expression of flagellar surface motility (Bloodgood and Workman, 1984), turn over rapidly in the intact flagellum (Bloodgood, 1984), and bind the lectin concanavalin A (Monk et al., 1983; Bloodgood et al., 1986). Immunolocalization studies using monoclonal antibodies to these HMW flagellar glycoproteins indicate that they are uniformly distributed along the entire flagellar surface (see Figure 6a,b, in Section III.C.1, below) in addition to being present on the rest of the cell's plasma membrane, albeit in a lower concentration (Bloodgood et al., 1986). Jarvik and Rosenbaum (1980) have described a paralyzed flagella mutant of *C. reinhardtii* that has a defect that affects both radial spoke structures in the axoneme and the mobility on acrylamide gels of the HMW flagellar membrane glycoproteins. Despite the obvious structural differences between the flagellar membrane and the rest of the cell's plasma membrane described above, there is as yet no clear evidence for any protein component unique to the flagellar membrane. In the case of *Euglena*, Rogalski and Bouck (1980) have described a glycoprotein they call "xyloglycorien" as unique to the flagellar surface and reservoir membrane. Unfortunately, it is very difficult to compare the flagellar membrane with the rest of the plasma membrane in *Chlamydomonas*; the cell body plasma membrane is closely apposed to the cell wall on the outside and the large, cup-shaped chloroplast on the inside. Despite the availability of cell-wall-less strains of *Chlamydomonas* (Davies and Plaskitt, 1971), no one has succeeded in obtaining a purified preparation of cell body plasma membrane, primarily because of the large, fragile chloroplast. However, it should be possible to perform surface-protein iodination using a cell-wall-less strain, deflagellate the cells, separate flagella from cell bodies, and then compare the two factions by acrylamide gel electrophoresis and autoradiography. For similar reasons, it has not been possible to directly compare the lipid composition of the flagellar membrane with that of the rest of the plasma membrane; Gealt et al. (1981) reported that flagella and deflagellated cell bodies had identical sterol compositions but differed in fatty acid composition. Dr. Aleksander Sikorski (University of Wroclaw, Poland, personal communication) performed spin-labeling experiments with 5-doxylstearate and found that vegetative flagellar membranes were much less fluid than gametic flagellar membranes.

The flagella play an important role in fertilization in *Chlamydomonas*, and gametic flagellar surfaces acquire certain unique properties during gametogenesis (Snell, 1985; van den Ende, 1985). Flagellar membranes of gametic cells contain species-specific, gamete-specific and mating-type-specific glycoproteins that function as sexual agglutinins and mediate the initial interactions between + and − gametes of the same species (Adair, 1985; van den Ende, 1985). Although present in small amounts, the sexual agglutinins have been purified and characterized for the + and − gametes of *C. reinhardtii* (Adair, 1985; Saito and Matsuda, 1984b) and the − gametes of *C. eugametos* (van den Ende, 1985). The agglutinin molecules are glycoproteins rich in hydroxyproline (Cooper et al., 1983), arabinose, and galactose (van den Ende, 1985). The agglutinins appear to be extrinsic membrane proteins that can be removed from the flagellar surface with ethylenediaminetetraacetate (EDTA) (Adair, 1985; Saito and Matsuda, 1984a) without killing the cell or disrupting the flagellar membrane. The agglutinins are of extremely high apparent molecular weight (Adair, 1985; van den Ende, 1985) and migrate much slower on SDS (sodium dodecyl sulfate)–acrylamide gels than do the 350-kd glycoproteins that are the principal protein components of both vegetative and gametic flagellar membranes.

III. DYNAMIC PROPERTIES OF THE *CHLAMYDOMONAS* FLAGELLAR SURFACE

Plasma membranes are extremely dynamic. The dynamics can be viewed at many levels, from the lateral diffusion and rotational movements of individual lipid and protein molecules up to the level of gross deformations of the plasma membrane involved in cell shape changes, cell motility, and endocytosis. The *Chlamydomonas* flagellar surface exhibits three principal kinds of temperature-dependent dynamic phenomena that are common to both vegetative and gametic cells: (1) turnover of membrane components in the intact flagellum; (2) rapid, bidirectional flagellar surface motility; and (3) global redistribution and shedding of particular classes of glycoproteins as visualized using lectins and antibodies. These phenomena appear to function in support of events in the mating process, for whole cell gliding motility, and during flagellar assembly and disassembly.

A. Turnover of Membrane Components

1. Vegative Cells

Treatment of cells with drugs that inhibit protein synthesis (cycloheximide) or protein glycosylation (tunicamycin) results in a reversible loss of flagellar surface adhesiveness and flagellar surface motility without any change in flagellar length or morphology (Bloodgood et al., 1979; Bloodgood, 1982). Cycloheximide treatment results in loss of the organized longitudinal array of intramembrane particles observed in the untreated flagellar membrane (Bloodgood, 1981). Removing the cells from either drug results in a gradual recovery of function at the flagellar surface. Pronase treatment of whole cells results in loss of flagellar surface adhesiveness and proteolytic modification of one of the HMW, surface-exposed glycoproteins (Bloodgood and May, 1983); cells can repair the pronase damage to the flagellar surface. Recovery from the effects of tunicamycin or pronase is prevented by cycloheximide (Bloodgood, 1982; Bloodgood and May, 1982), indicating that new protein synthesis, processing, and insertion into the flagellum is involved in the recovery from the effects of these drugs. Treatment of cells with cycloheximide at 4°C or less does not result in loss of flagellar surface adhesiveness or flagellar surface motility; this is interpreted as an inhibition of flagellar membrane turnover at low temperatures (Bloodgood et al., 1979). These observations suggest that protein turnover is occurring in the intact flagellum and that some flagellar glycoprotein, which turns over relatively rapidly within the flagellum, is associated with flagellar surface adhesiveness and flagellar surface motility. Radioactive pulse-labeling studies have directly demonstrated turnover of flagellar membrane proteins in the intact flagellum (Bloodgood, 1984; Remillard and Witman, 1982) and have shown that membrane proteins turn over much faster than axonemal proteins.

In many cell types, plasma membrane turnover is brought about by endocytosis resulting in the formation of endosomes, which are then delivered to the Golgi compartment. This process cannot occur in the flagellum because the flagellar membrane is closely opposed to the underlying cytoskeleton (the axoneme) and vesicles are never seen within the flagellum. Indeed, incubation of cells in soluble horseradish peroxidase followed by cytochemical demonstration of peroxidase activity provided no evidence for endocytosis in *Chlamydomonas* (E. N. Levin and R. A. Bloodgood, unpublished results). Instead, the flagellar membrane appears to continuously release small membrane vesicles into the medium (Bergman et al., 1975; McLean et al., 1974; Snell, 1976); these membrane vesicles still possess the fuzzy coat (glycocalyx), and acrylamide gel electrophoresis shows them to contain the HMW membrane glycoproteins. This membrane release, of necessity, must promote turnover of lipid and protein components of the flagellar membrane. The machinery for protein synthesis is absent from the flagella, and all protein synthesis occurs within the cell body. Little is currently known about the pathways for inserting newly synthesized proteins into the flagellum. The basal body complex can be viewed as a plug at the base of the flagellum; although ribosomes are excluded from entry into the flagellum, it is often assumed that soluble proteins synthesized on free polysomes and released into the cell body cytoplasm can gain ready access to the interior of the flagellum. This may not necessarily be the case; recent evidence suggests that flagellar protein complexes can be preassembled in the cytoplasm (Mitchell and Rosenbaum, 1985b). The possibility exists that entry into the flagellum of individual proteins or protein complexes may have to be facilitated by some as yet unknown mechanism; this would allow some selectivity in terms of which proteins within the cytoplasm are admitted into the "flagelloplasm." Entry of flagellar membrane glycoproteins into the flagellar membrane presumably occurs through a vesicular pathway, as has been demonstrated for most other plasma membrane glycoproteins. The problem that arises here is that membrane vesicles in transit from the Golgi apparatus cannot enter the flagellum and then insert their contents by direct exocytosis into the existing flagellar membrane. Presumably, the cytoplasmic membrane vesicles carrying the flagellar glycoproteins must fuse with the cell body plasma membrane, after which lateral diffusion within the plasma membrane carries the glycoproteins into the flagellar membrane, where they could be stabilized and concentrated by interactions with the underlying cytoskeleton. One potential problem with this scenario is the existence of the flagellar necklace and the flagellar bracelet (Weiss et al., 1977), which might limit exchange by diffusion between the two plasma membrane compartments (flagellar membrane vs. all the rest of the plasma membrane). One of the most highly developed ciliary necklace-type structures exists in the connecting cilium of the rod photoreceptor cell (Rohlich, 1975), despite the fact that enormous amounts of opsin protein must move through this membrane region each day to sustain the turnover of the disk membranes of the rod outer segments. During ciliary regeneration in *Tetrahymena*, the ciliary necklace remains intact (Satir et al., 1976). These observations suggest that the ciliary/flagellar necklace structure may not seriously impede the mobility of proteins in the plane of the plasma membrane. Dentler (1981) provides an excellent discussion of the possible roles for the ciliary/flagellar necklace and the issue of the flagellar membrane as a separate compartment.

2. Gametes and Mating

In the case of gametic cells, membrane vesicles released from the flagella into the medium have been referred to as "gamone" and possess the flagellar membrane sexual agglutinin molecules that interact with flagellar surfaces of cells of the opposite mating type (Bergman et al., 1975; Snell, 1976; McLean et al., 1974). This material has the capacity to isoagglutinate cells of the opposite mating type; in fact, the early events of mating can be stimulated by mixing live intact gametes of one mating type with fixed cells, isolated flagella, flagellar membranes, or purified agglutinin from cells of the opposite mating type (Mesland and van den Ende, 1979; Snell and Roseman, 1979; Saito and Matsuda, 1984b). Although it is not known whether gametic flagellar membranes are turning over at the same rate or slower than vegetative flagellar membranes, it is clear that gametic flagellar interaction during the early stages of the mating process results in a greatly increased rate of turnover, at least for some of the flagellar membrane components including the sexual agglutinin glycoprotein molecules. Cell bodies of gametic cells of *C. reinhardtii* (Saito et al., 1985) and *C. eugametos* (Pijst et al., 1983) contain a large pool (25–100 times the amount in the flagella) of the flagellar agglutinin glycoprotein. In *C. eugametos*, this reserve of agglutinin molecule appears to be associated with the cell body plasma membrane (Pijst et al., 1983), whereas it seems to be in a "soluble" form in the cell bodies of *C. reinhardtii* gametes (Saito et al., 1985); as the cell body pool of sexual agglutinin glycoprotein molecules enters the flagella, the agglutinin molecules are successively inactivated during normal gametic flagellar interaction by an as yet undefined mechanism.

Flagellar agglutinin activity can be removed from the flagellar surface of *C. reinhardtii* by EDTA or trypsin treatment; the flagella will regain agglutinin activity, presumably by flagellar membrane protein turnover, and this recovery is inhibited by cycloheximide in both mating types and by tunicamycin in mating type + gametes (Solter and Gibor, 1978; Saito and Matsuda, 1984a; Saito et al., 1985). EDTA and trypsin treatments appear to remove both the flagellar and the cell body-associated agglutinin molecules, presumably by speeding up flagellar turnover (Saito et al., 1985) and driving the cell body agglutinins into the flagellar membrane, where they are exposed to the action of the EDTA or trypsin. The ability of researchers to strip intact agglutinin molecules from live cells by treatment with EDTA suggests that the agglutinin molecules are not intrinsic membrane proteins with a portion of their protein core embedded in the lipid bilayer (Saito and Matsuda, 1984a; Adair, 1985). This raises the possibility that there is some population of integral flagellar membrane proteins that serve as a receptor or anchor for the agglutinin molecules.

In unmated gametes, inhibition of protein synthesis with cycloheximide or protein glycosylation with tunicamycin results in little or no loss of agglutinin activity from the flagellar surface of *C. reinhardtii* (Saito et al., 1985; Snell and Moore, 1980), *C. Eugametos* (Pijst, 1985), and *C. moewusii* (Wiese et al., 1984). This could result from the large size of the agglutinin pool in the cell body or from a very slow turnover of flagellar agglutinin; the latter is probably the case since Saito et al. (1985) showed that the half-life of the cell body agglutinins is 12 hours in unmated gametes vs. 15 minutes after initiation of mating. Interaction of + and − gametes results, in both *C. reinhardtii* and *C. moewusii*, in an inactivation of the flagellar surface-exposed agglutinins (Pijst et al., 1984; Snell and Roseman, 1979), a major stimulation of the rate of turnover of the flagellar agglutinins (Saito et al., 1985; Snell and Moore, 1980) and presumably an increase in the rate of synthesis of the agglutinin. While Saito et al. (1985) presented data arguing that the entire cell-body pool of agglutinin in *C. reinhardtii* was able to enter the flagellum and thereby be subjected to destruction by trypsin or extraction with EDTA, Pijst (1985) presented data for *C. eugametos* showing that cycloheximide resulted in loss of agglutinin from the flagellum without depletion of the cell-body. Since Pijst et al. (1983) had demonstrated that at least some of the cell-body pool of agglutinin was in the plasma membrane, Pijst (1985) proposed that there may be a barrier separating the flagellar membrane compartment from the rest of the cell's plasma membrane. When added at the time of mixing + and − gametes, inhibitors of protein synthesis (Snell and Moore, 1980), protein glycosylation (Snell, 1981), and proline hydroxylation (Cooper et al., 1983) result in loss of flagellar adhesiveness between *C. reinhardtii* gametes; this is presumed to result from protein turnover exhausting the supply of functional agglutinin both within the flagellum and the cellular pool. Presumably, agglutinin molecules lacking the normal posttranslational modifications (glycosylation or proline hydroxylation) are either inactive or fail to be inserted into the flagellar membrane or both. For convenience, these studies were performed with a mutant of *C. reinhardtii* (imp-1) that is defective in the normal de-adhesion mechanism and continues to exhibit flagellar agglutination for hours in the absence of any inhibitors. Snell and Moore (1980) reported on the changes in agglutinin pool size that occur with time after the initiation of mating in *C. reinhardtii.*

After cell fusion (heterokaryon formation) is achieved during mating in *Chlamydomonas*, the adherent flagella of the two gametes de-adhere (Snell and Roseman, 1979). Almost nothing is known about the mechanism of normal flagellar de-adhesion, but it would not be surprising if it involved turnover of flagellar membrane components, either the loss of some component from the flagellar surfaces essential for maintaining contact or the insertion of something into the flagellar surface that actively promotes de-adhesion.

Snell et al. (1983) observed that gametic agglutination induced the synthesis and insertion into the flagellar membrane of two proteins of 220 and 165 kd. These proteins are exposed at the flagellar surface but do not appear to possess any agglutinin activity in a bioassay (Snell, 1985); their function remains unknown.

B. Flagellar Surface Motility

One of the most remarkable phenomena associated with the *Chlamydomonas* flagellar surface is referred to as flagellar surface motility and is exhibited as the gliding motility of the cell along a solid substrate (Figure 3) or can be visualized artificially by the movement of polystyrene microspheres along the flagellar surface (Figure 4). A number of observations have also led certain workers to feel that the energy-transduction system powering gliding motility may also be involved in the early events of mating, particularly for use during flagellar alignment.

1. Whole-Cell Gliding Motility

Lewin (1952) observed that cells of *C. moewusii* could "creep" along a solid substrate and that this whole-cell locomotion was dependent upon the flagella. He further showed that paralyzed flagella mutants, in which the flagella could not propagate bends, exhibited this "creeping" or gliding motility (Lewin, 1954). Lewin (1952) postulated that the motor responsible for gliding motility was also utilized for flagellar reorientation during mating. Bloodgood (1981) went on to characterize gliding motility in *C. reinhardtii* in detail and to postulate that gliding motility and microsphere movement were two manifestations of the same motor. Lewin (1982) reported on the isolation of nongliding mutant strains of *C. moewusii* that also failed to exhibit microsphere motility. These mutants, more fully characterized by Reinhart and Bloodgood (1984), constitute a genetic argument that gliding motility and microsphere movements on the flagellar surface are two expressions of the same basic energy-transducing system. The expression of gliding motility usually occurs after the cells have adopted a characteristic gliding morphology in which the two flagella are oriented at 180° to one another (Figure 3). Gliding occurs at approximately 2 μm/s in the directions defined by the flagellar orientation; cells can reverse direction. A variety of observations (Bloodgood, 1981) suggest that, at any particular time, it is the leading flagellum that is providing the force for gliding. Polystyrene microspheres can be observed to exhibit saltatory movements along the free

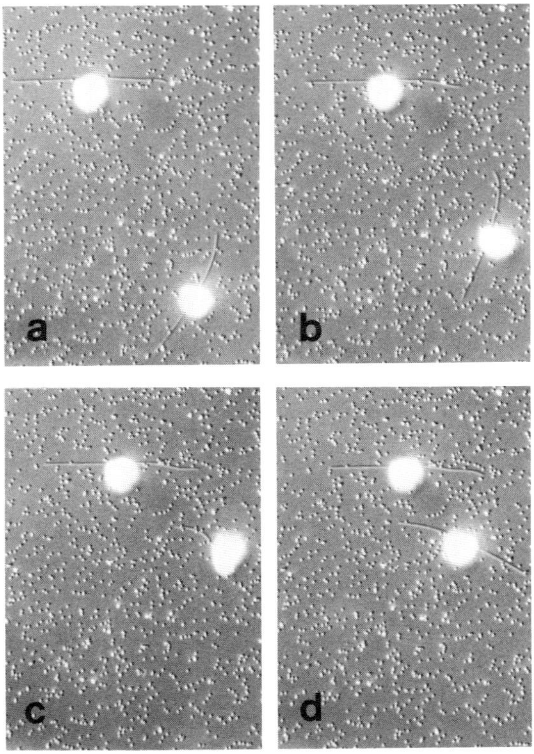

Figure 3. Series of 35-mm frames showing *Chlamydomonas moewusii* strain M475 cells gliding along the surface of a glass slide that has a number of polystyrene microspheres attached to it. The path taken by the cells is dictated by the orientation of the leading flagellum. (Micrographs courtesy of Mr. Frederick Reinhart.) Differential interference contrast optics. ×1,000.

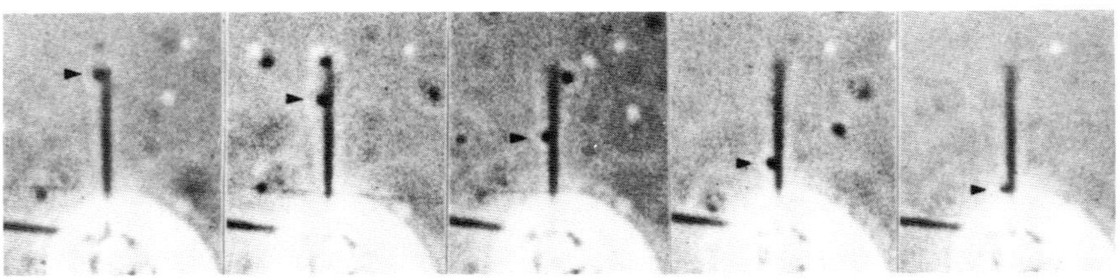

Figure 4. Sequence of frames from a 16-mm film showing the movement of 0.357-μm-diameter polystyrene microspheres along the flagellar surface of a *C. reinhardtii* strain pf-18 cell. The interval between adjacent micrographs is 0.9 sec. Phase contrast optics. (Reproduced from Bloodgood, 1977.) ×1,827.

surface of a flagellum that is involved in gliding; the movements of the microspheres are seen to be independent of the direction of gliding, which supports the argument that local domains of the flagellar membrane are independent of one another. B. Daniels and C. King at University College London (unpublished results) used interference reflection microscopy to study the flagella-substrate interactions during gliding motility of *C. reinhardtii*; they observed multiple sites of close contact on each flagellum during gliding. Further, these sites were observed to migrate along the flagella during gliding. Gliding motility appears to be an alternative form of whole-cell locomotion (flagella-based swimming motility being the other) and a normal physiological expression of flagellar surface motility.

2. *Polystyrene Microsphere Movements*

Bloodgood (1977) observed that polystyrene microspheres (0.28–0.85 μm in diameter) adhered to and were transported along the surface of the flagella of vegetative and gametic cells of both wild-type and nonmotile mutants of *C. reinhardtii* and *C. moewusii* (Figure 4). The microspheres were transported at 2 μm/s in a bidirectional, saltatory manner; movement always occurred parallel to the long axis of the flagellum, as if the microspheres were constrained to defined "tracks" within the flagellar membrane. Increasing the size of the microspheres or increasing the viscosity of the medium had no effect on the velocity of microsphere movement. The nature of the object being carried is not specific; microspheres derivatized in a variety of ways, and even cells of a nonmotile strain of *Escherichia coli* will be moved along the flagellar surface (Bloodgood, 1977). Movement does not occur on dead cells or isolated flagella. Using cell wall-less mutants, he was able to demonstrate that microsphere movement does not occur along the rest of the cell's plasma membrane; when microspheres reach the base of the flagella, they do not move onto neighboring areas of plasma membrane, perhaps restricted by the intramembrane particle specializations (flagellar necklace and bracelet) located at that site or by the lack of an appropriate cytoskeleton underlying that region of the plasma membrane.

An important characteristic of microsphere movement is that it is local; a number of microspheres can be observed moving independently on the same flagellum at the same time. This suggests that there exist small, independent domains within the flagellar membrane; the microsphere movements cannot be due to bulk membrane or lipid flow (such as might be created by continuous turnover of the flagellar membrane due to insertion at one end and pinching off of membrane into the medium at the other). Every microsphere that becomes mechanically associated with the flagellar surface exhibits periods of movement. Although the flagellar surface cannot be said to possess specific receptors for the microspheres, it is clear that the entire flagellar surface possesses exposed sites for microsphere binding that arc coupled to a force-transducing system. Perhaps a system evolved for whole-cell locomotion along a variety of solid and semisolid substrates would not be expected to possess surface-binding sites with a great deal of specificity.

Methods have been designed for quantitating microsphere adhesion to and movement along the flagellar surface (Bloodgood et al., 1979; Hoffman and Goodenough, 1980). Using these methods, it has been possible to identify a variety of treatments that reversibly inhibit flagellar surface adhesion and/or flagellar surface motility (Bloodgood et al., 1979; Hoffman and Goodenough, 1980; Snell et al., 1982; Detmers and Condeelis, 1986). Table 1 provides a comparison of the effect of various experimental treatments on microsphere adhesion, microsphere movement, and mating efficiency of gametes in *C. reinhardtii.* Note that a number of the conditions that inhibit microsphere movement also inhibit the mating of *Chlamydomonas* gametes. These inhibitors are all reversible and can be classified into those that act quickly (minutes at most) vs. those that act much more slowly (hours). The latter class includes cycloheximide and tunicamycin and is thought to act through a normal process of protein turnover (see Section III.A.1, above) which results in depleting the flagellar membrane of a critical protein (in the case of cycloheximide) or in replacing a population of glycoproteins with nonglycosylated counterparts (in the case of tunicamycin). It is interesting that cycloheximide results in both an inhibition of microsphere movement (Bloodgood et al., 1979) and a disorganization of the longitudinally oriented intramembrane particle arrays in the flagellar membrane (Bloodgood, 1981). Note in Table 1 that a number of the treatments that inhibit flagellar adhesiveness and/or flagellar surface motility also reduce the mating efficiency of gametes by affecting one or more of the steps in the fertilization process.

Several observations suggest that calcium is involved in flagellar surface motility: (1) Reduction of the free-calcium concentration in the medium below the micromolar level results in a reversible inhibition of microsphere movement (Bloodgood et al., 1979). (2) The anesthetic lidocaine (Xylocaine) reversibly inhibits microsphere movement (Snell et al., 1982). Lidocaine is known to interfere with divalent cation movement across membranes and to release membrane-bound calcium (Low et al., 1979; Seeman, 1972). (3) Trifluoperazine reversibly inhibits flagellar surface motility (Detmers and Condeelis, 1986). Trifluoperazine has been shown to specifically bind to the calcium regulatory protein calmodulin (Levin and Weiss, 1979), which is known to be abundant in the *Chlamydomonas* flagellum (Gitelman and Witman, 1980).

Rapid, local, bidirectional surface motility is not unique to *Chlamydomonas* flagella; similar phenomena visualized by the movement of polystyrene microspheres have been observed on the cilia of sea urchin

Table 1. Reversible Inhibitors of Flagellar Surface Adhesion, Flagellar Surface Motility, and Mating in *Chlamydomonas reinhardtii*

Inhibitor or Condition	*Microsphere Adhesion*	*Microsphere Movement*	*Mating of Gametes*	*Reference*
4°C	Unaffected	Inhibited	Inhibited	Bloodgood et al., 1979; Goodenough et al., 1980
Methylxanthines		Inhibited		Bloodgood et al., 1979
Lidocaine (Xylocaine)		Inhibited	Inhibited	Snell et al., 1982
Low calcium (less than 1 μm)	Unaffected	Partial inhibition	Unaffected (except rate)	Bloodgood and Levin, 1983; Bloodgood et al., 1979; Goodenough et al., 1980
Trifluoperazine		Inhibited	Inhibited	Detmers and Condeelis, 1986
Colchicine	Unaffected	Unaffected		Bloodgood, 1977;
	Inhibited	Unaffected	Inhibited	Hoffman and Goodenough, 1980
Cytochalasin	Unaffected	Unaffected		Bloodgood et al., 1979;
	Inhibited		Inhibited	Hoffman and Goodenough, 1980; Detmers et al., 1983
Cycloheximide	Inhibited	Inhibited		Bloodgood et al., 1979
Tunicamycin	Inhibited	Inhibited		Bloodgood, 1982
Pronase	Inhibited	Unaffected	Inhibited	Bloodgood and May, 1982; Wiese and Hayward, 1972
Chymotrypsin	Inhibited	Unaffected	Inhibited	Hoffman and Goodenough, 1980
Trypsin	Unaffected	Inhibited	Inhibited	Hoffman and Goodenough, 1980; Solter and Gibor, 1978; Wiese and Hayward, 1972

embryos (Bloodgood, 1980), axopodia of Heliozoan protozoa (Bloodgood, 1978; Bowser and Bloodgood, 1984; Kanno and Ishii, 1979; Suzaki and Shigenaka, 1982; Troyer, 1975) and the reticulopodia of Foraminiferan protozoa (Bowser and Bloodgood, 1984; Bowser et al., 1984; Bowser and Rieder, 1985). All these systems involve long cellular extensions filled with a microtubule-based cytoskeleton; it has been clearly shown that the polystyrene microspheres moving on the surface of *Allogromia* reticulipodia follow the paths of microtubules in the underlying cytoplasm (Bowser and Rieder, 1985). In the case of Heliozoan and Foraminiferan protozoa, rapid surface motility is utilized for transport of food organisms along the cell surface (Bowser et al., 1985; Suzaki et al., 1980).

As mentioned in Section III.B.1, above, whole-cell gliding motility and microsphere movement appear to be two different manifestations of the same energy-transducing system. Both occur at the same velocity in a bidirectional manner, and mutants defective in gliding motility fail to exhibit microsphere movement along the flagellar surface (Lewin, 1982; Reinhart and Bloodgood, 1984).

3. Relationship of Flagellar Surface Motility to the Early Events of Mating

The flagella surfaces of mating type + and − gametes are the initial sites of interactions between gametes during fertilization in *Chlamydomonas.* Following the initial adhesion of the flagellar surfaces, the pairs of adherent flagella undergo a process of reorientation (migration of the sites of contact) whereby the flagellar tips are put into register and "locked" into this position (Bloodgood, 1982; Mesland, 1976; Mesland et al., 1980; Hoffman and Goodenough, 1980); this flagellar orientation is a necessary condition for the later events in mating: flagellar signaling, cell-wall release, mating structure activation, and cell fusion (Snell, 1985). It has been suggested by Lewin (1952), by Hoffman and Goodenough (1980), and by Bloodgood and colleagues (1979; see also Bloodgood, 1982) that the flagellar surface motility phenomenon expressed as gliding motility and microsphere movement is also responsible for this flagellar realignment process during mating. Gametic cells exhibit polystyrene microsphere movements (Bloodgood, 1977; Hoffman and Goodenough, 1980; Snell et al., 1982). It is interesting that certain drugs (e.g. lidocaine or trifluoperazine) that reversibly inhibit microsphere movement also reversibly inhibit the early events of mating (but not the initial flagellar adhesion) (Snell, 1985; Detmers and Condeelis, 1986). Table 1 lists a number of treatments that interfere with both flagellar surface motility and successful mating. Hoffman and Goodenough (1980) observed that, upon the completion of flagellar alignment (the tip-locked stage), microspheres on the flagellar surface would migrate to and remain at the flagellar tips until flagellar de-adhesion occurred. Since Hoffman and

Goodenough (1980) also observed that the movement of contact points during flagellar realignment during mating is slower than the movement of polystyrene microspheres on the same flagella, then it is possible that flagellar tip alignment may instead be achieved by a process of slow, bulk redistribution of flagellar membrane glycoproteins more akin to the process to be described below in Section III.C.5.

4. *Flagellar Surface Motility and Flagellar Assembly*

Whole-cell gliding motility and flagellar orientation during mating are expressions of flagellar surface motility specific to *Chlamydomonas.* However, other systems, such as the cilia on sea urchin blastulae (Bloodgood, 1980), exhibit similar forms of surface motility that must serve other ends. In *Chlamydomonas*, it is known that polystyrene microsphere motility occurs on flagella during their assembly (regeneration after deflagellation) (Bloodgood, 1977). It has been shown that growth of at least the outer doublet microtubules in flagella occurs at the distal end of the growing structures (Rosenbaum and Child, 1967; Witman, 1975); recently, Mitchell and Rosenbaum (1985b) have shown that aggregates of dynein polypeptides are preassembled in the cell body before insertion into the flagellum. The suggestion was originally made by Dr. Joel L. Rosenbaum (personal communication) that flagellar surface motility as visualized on the outside surface of the flagellar membrane may actually represent a transport system for movement of flagellar axoneme precursors along the flagellum either inserted into or attached to the inner surface of the flagellar membrane. One might wish initially to discard this notion because flagellar surface motility (at least in vegetative cells) does not show any bias in directionality; approximately the same number of microsphere saltations occur in the distal as in the proximal direction along the flagellar surface (Bloodgood, 1977). Nevertheless, such a surface motility system could still serve to facilitate the distribution of flagellar axoneme subunits along the length of the flagellum beyond what could be provided by simple diffusion. This process would be particularly effective for large protein aggregates with low diffusion constants, especially considering that the interior of the flagellum is congested with axonemal structures. This idea may have implications for our understanding of regeneration kinetics and flagellar length regulation in *Chlamydomonas*, something that is unlikely to be related to simple diffusion of subunits from the cell body (Jarvik et al., 1984).

5. *Cellular Mechanism for Flagellar Surface Motility*

Flagellar-surface motility is hypothesized to result from the movements of flagellar membrane components within the plane of the flagellar membrane. In particular, it is hypothesized that flagellar surface motility is associated with transmembrane glycoproteins that possess at least three domains: (1) a surface-exposed domain containing sites for substrate interaction; (2) a transmembrane domain (probably hydrophobic and α-helical); and (3) a cytoplasmic domain with sites that interact with an energy-transducing system, which may be anchored on the outer doublet microtubules (Figure 5). These proteins, individually or in small groups, are presumed to be moved laterally in the plane of the plasma membrane in an energy-dependent manner. The flagellar protein fulfilling the role of the transmembrane component of the system and the nature of the energy-transducing system responsible for exerting force on this transmembrane component have not been unequivocally demonstrated.

The composition of the *C. reinhardtii* flagellar membrane is surprisingly simple and is dominated by a HMW glycoprotein that migrates with an apparent molecular weight around 350 kd (Bergman et al. 1975; Bloodgood and May, 1982; Snell, 1976; Witman et al., 1972; Bloodgood et al., 1986) that can be resolved into two components (Bloodgood and May, 1982; Bloodgood, 1984). These HMW flagellar glycoproteins possess multiple

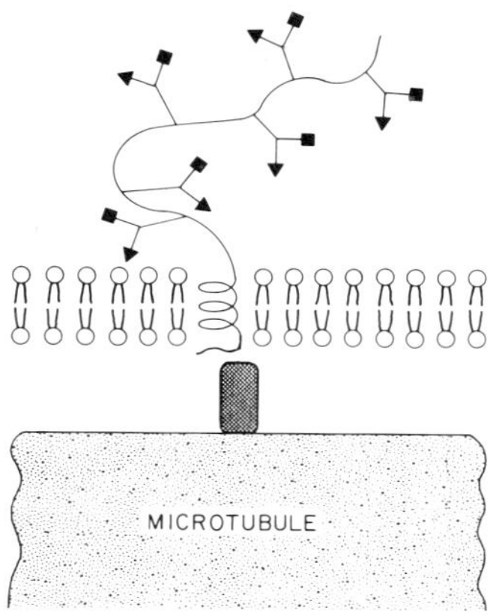

Figure 5. Diagram illustrating a model for the organization of the flagellar surface of *C. reinhardtii.* The HMW flagellar glycoprotein molecule is shown to have a large extracellular domain possessing a number of glycosylation sites, each possessing epitopes for the binding of concanavalin A (squares) and the FMG-1 anticarbohydrate monoclonal antibody (triangles). In addition, the molecule is shown to have a transmembrane α-helical portion and a cytoplasmic domain that connects with a structure that provides a membrane-axoneme linkage and may contain the energy-transducing system that powers flagellar surface motility. The linking structure is anchored on an outer doublet microtubule. (Modified from a diagram in Bloodgood et al., 1986.)

sites of glycosylation, all of which contain epitopes recognized by the lectin concanavalin A and an anticarbohydrate monoclonal antibody (Bloodgood et al., 1986). Several additional minor flagellar membrane proteins have been described that are common to the vegetative and gametic flagellar membranes of *C. reinhardtii* (Snell, 1976; Remillard and Witman, 1982; Bloodgood, 1974; Bloodgood and Workman, 1974; Bloodgood et al., 1986). In addition, gametic flagellar membranes possess specific agglutinin molecules, which are very HMW glycoproteins present in very low amounts (Adair, 1985; van den Ende, 1985; Saito and Matsuda, 1984b).

Several lines of evidence suggest that the 350-kd HMW glycoproteins common to the vegetative and gametic flagellar membranes of *C. reinhardtii* serve the role of the membrane component that both binds to an external substrate (be it a planar surface, a polystyrene microsphere, or another flagellar surface) and mechanically couples that substrate to a force-transducing system:

1. These proteins are distributed over the entire surface of the flagellum (Bloodgood and May, 1982; Bloodgood et al., 1986).
2. Inhibition of protein synthesis results in loss of flagellar surface adhesiveness and flagellar surface motility (Bloodgood et al., 1979; Bloodgood, 1982); these glycoproteins have been shown to turn over rapidly in the intact flagellum (Bloodgood, 1974), and this turnover correlates with the loss of flagellar surface function.
3. Pronase digestion of the flagellar surface results in loss of microsphere binding; the principal components that are modified by this proteolytic treatment are these HMW glycoproteins (Bloodgood and May, 1982).
4. Use of a substrate immobilized iodination system results in labeling primarily the HMW glycoproteins (Bloodgood and Workman, 1984).
5. Treatment of cells with polyclonal (Bloodgood and May, 1982) and monoclonal antibodies (Bloodgood et al., 1986) recognizing these HMW glycoproteins results in an inhibition of polystyrene microsphere movements. This observation may also suggest that these glycoproteins must normally be allowed to move laterally in the plane of the flagellar membrane to allow expression of flagellar surface motility.

It should be emphasized that the information described above applies only to *C. reinhardtii.* The flagellar surface of another species, *C. moewusii,* differs greatly in its membrane protein composition, yet both species exhibit gliding motility and microsphere movement with the same characteristics (Reinhart and Bloodgood, 1984). A large number of monoclonal antibodies to *C. reinhardtii* flagellar membrane components fail to react with *C. moewusii* (Reinhart and Bloodgood, unpublished results), and a number of monoclonal antibodies prepared to the *C. Eugametos* flagellar membrane react with *Chlamydomonas moewusii* but fail to react with *C. reinhardtii* (H. van den Ende, unpublished results). The flagellar surfaces of different species of *Chlamydomonas* also differ in a number of other properties, including glycoprotein composition and sensitivity to pronase (Reinhart and Bloodgood, unpublished results).

Probably the best way to convincingly demonstrate that the HMW glycoproteins are playing an important role in flagellar surface motility will be to obtain mutant cell lines with defects in defined HMW flagellar membrane glycoproteins. As a preliminary step toward this, a panel of mouse hybridomas secreting monoclonal antibodies recognizing carbohydrate and protein epitopes on HMW flagellar membrane proteins in vegetative *C. reinhardtii* has been obtained (Bloodgood et al., 1986); most other efforts have been directed toward obtaining monoclonal antibodies recognizing gamete specific flagellar proteins (Adair, 1985; Snell et al., 1985). The monoclonal antibodies recognizing the principal flagellar glycoproteins (Bloodgood et al., 1986) are being utilized to interfere with flagellar surface functions and to select for mutant cell lines with defects in these proteins by utilizing fluorescent-activated cell sorting.

Although much has been learned about the flagellar membrane glycoproteins that may provide the adhesion and transmembrane coupling functions necessary for the expression of flagellar surface motility, much less is currently known about the energy-transducing system itself. *C. reinhardtii* flagella contain a 3.0-S calcium-specific ATPase (Watanabe and Flavin, 1976) that has been localized to the membrane-matrix compartment of the flagellum (Bessen et al., 1980). Gitelman and Witman (1980) have reported the presence of the calcium regulatory protein calmodulin in nonionic detergent extracts of *Chlamydomonas* flagella. Since flagellar surface motility is known to be calcium dependent (Bloodgood et al., 1979), one can imagine that a calmodulin-Ca-ATPase system localized in the slender bridges observed in electron micrographs to connect the membrane with the axoneme (Figures 2A–C and 5) could be responsible for the force generation necessary to rapidly move glycoprotein molecules in the plane of the flagellar membrane.

The possibility remains that axonemal dynein ATPases could somehow be utilized for energy transduction at the flagellar surface; this possibility appears to be ruled out by the observation made by D. Mitchell and J. L. Rosenbaum (unpublished observations) that a mutant strain of *C. reinhardtii* lacking outer dynein arms exhibited normal polystyrene microsphere movements along the flagellar surface.

It is possible that the mutant cell lines of *C. reinhardtii* defective in gliding motility and microsphere movements (Lewin, 1982) will be useful for understanding the energy-transduction system operating at the flagellar surface. Although some of the nongliding mutants appear to have defects at the flagellar surface (Reinhart and Bloodgood, 1984), others appear to have a normal flagellar surface and hence are prime candidates for possessing defects in the energy-transduction system.

C. Redistribution of Flagellar Membrane Proteins

1. Vegetative Cells

Goodenough and Jurivich (1978) reported that antibodies generated to whole flagella from *C. reinhardtii* bind to the flagellar surface and redistribute to the flagellar tip of gametic cells; they termed this phenomenon "tipping" in analogy to the "capping" phenomenon originally described in lymphocytes (dePetris, 1977; Bourguignon and Bourguignon, 1984). Bloodgood et al. (1986) have recently shown that two probes (the lectin concanavalin A and a monoclonal antibody designated FMG-1) that bind to carbohydrate groups associated with the HMW flagellar membrane glycoproteins exhibit a dramatic temperature-dependent redistribution along and shedding from the vegetative flagellar surface of *C. reinhardtii* strain pf-18 cells. When glutaraldehyde-fixed cells or live cells at 4°C are labeled with concanavalin A or the FMG-1 anticarbohydrate monoclonal antibody, the flagellar surfaces and the cell wall label uniformly (Figure 6). When the cell wall is not present (after mating of gametic cells or by use of a cell wall-less strain), the plasma membrane surrounding the cell body is also observed to stain. There are multiple sites for binding of the lectin and antibody on each HMW glycoprotein molecule; immunoaffinity binding studies have shown that all of the major peptides generated by V-8 protease digestion bind both the lectin and the antibody, although the two ligands do not compete for binding to the same epitopes. If live cells that are labeled in the cold are warmed to 24°C, the labeled lectin or antibody is observed to redistribute along the flagellar surface in a characteristic manner (Figure 7). The label initially clears away from the base of the flagella (Figure 7b,g) until it is entirely aggregated at the flagellar tips (Figure 7c,h), after which the entire aggregate of label migrates to the bases of the flagella (Figure 7d,i). The label is subsequently shed from the flagellar base yielding cells with unlabeled flagella (Figure 7e,j); the complete process takes between 5 and 15 minutes at 24°C. The same pattern of redistribution is observed whether directly labeled monoclonal antibody is followed in the fluorescence microscope on a single live cell with time or aliquots from a population of cells are fixed and labeled with fluorescein isothiocyanate(FITC)-conjugated second antibody at various times after warming. During the flagellar membrane redistribution process, which takes less than 10 minutes and is highly temperature dependent, there is no change in the intensity or distribution of label associated with the cell wall (Figure 7). The length and general

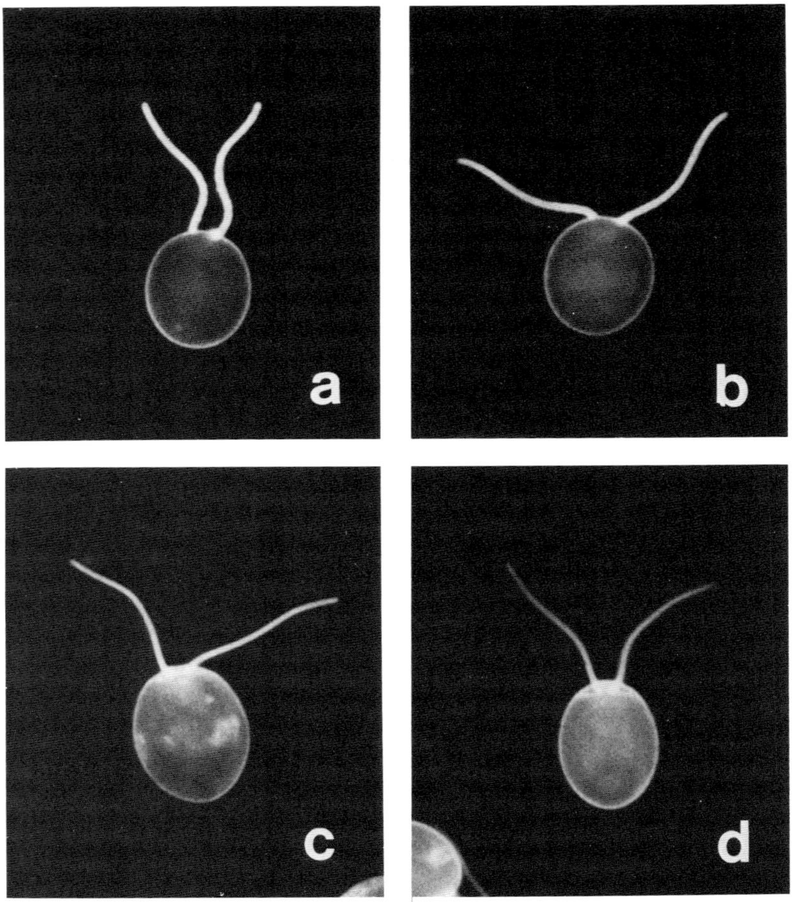

Figure 6. Figures illustrating the distribution of cell-surface-exposed binding sites on *C. reinhardtii* strain E2 cells for (A, B) an anticarbohydrate monoclonal antibody (FMG-1) and (C, D) the lectin concanavalin A. In each case, both the flagellar surface and the cell wall are uniformly labeled. The micrographs were taken using an epifluorescence microscope. The concanavalin A was directly labeled with FITC, whereas the monoclonal antibody was visualized by indirect immunofluorescence using a commercially available FITC-labeled goat anti-mouse IgG (immunoglobulin G) antibody. ×2,000.

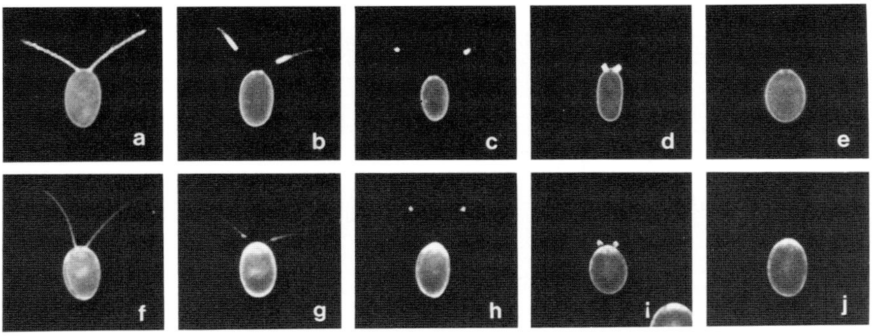

Figure 7. Redistribution of the FMG-1 anticarbohydrate monoclonal antibody (a–e) and FITC-concanavalin A (f–j) along the flagellar surface of *C. reinhardtii* strain pf-18 cells. In each case, the live cells were incubated with the ligand at 4°C, washed at 4°C, and then warmed to 24°C. At various times after warming, cells were fixed with glutaraldehyde; in the case of the monoclonal antibody, cells were then labeled with an FITC-conjugated second antibody prior to photographing in an epifluorescence microscope. Each series of frames represents the temporal sequence of stages in the redistribution process. (Reproduced from Bloodgood et al., 1986.) All, ×816.

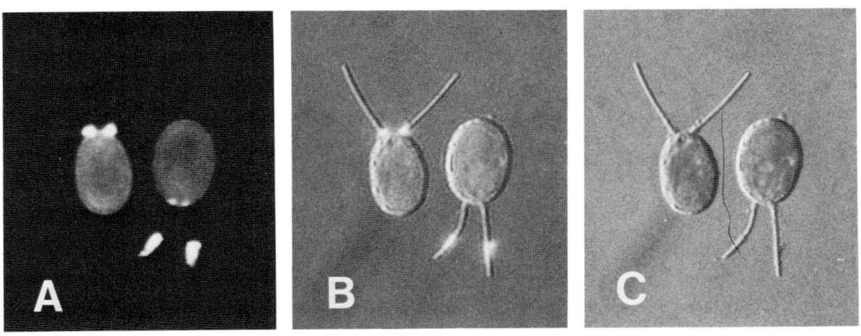

Figure 8. Comparison of the fluorescence and differential interference contrast images of two cells exhibiting different stages of redistribution of the FMG-1 anticarbohydrate monoclonal antibody. Panel A was photographed using epifluorescence optics. Panel C was photographed using differential interference contrast optics. Panel B was photographed using both fluorescence and DIC optical systems simultaneously. (Reproduced from Bloodgood et al., 1986.) All, ×1,440.

morphology of the flagella appear to be unchanged during this process of redistribution (Figure 8); studies that involve a second round of labeling with the antibody or lectin demonstrate that new binding sites appear within the cleared regions of the flagella (Bloodgood et al., 1986). Since concanavalin A and the monoclonal antibody do not compete for binding to the same epitope, it is possible to directly determine if the HMW glycoproteins are redistributing along with the labeled antibody. Redistribution is induced with unlabeled antibody; at various times, parallel samples are fixed. One sample from each time point is labeled with FITC-second antibody to localize the monoclonal antibody, while the other sample from each time point is labeled with FITC-concanavalin A in order to localize the HMW flagellar glycoproteins. The pattern of redistribution of label with time is very similar in the two samples, indicating that the antibody is redistributing in association with the glycoproteins it recognizes (Bloodgood et al., 1986). This important fact has not been directly demonstrated in most other cases where redistribution of lectins and antibodies along cell surfaces has been reported (dePetris, 1977).

It is unclear whether the antibody or lectin actually induces the redistribution that is observed or simply serves as a tag allowing the observer to visualize a process of membrane flow or glycoprotein movement that occurs normally in the absence of the antibody or lectin. Since the flagellar surface redistribution phenomenon in many ways resembles the "capping" phenomenon described in lymphocytes and many other cell types, one might conclude that it is a response to the cross-linking of the flagellar glycoproteins by these multivalent ligands. It has not been possible to determine an answer to this question since the noncross-linking analogs (succinyl-concanavalin A and the Fab′ fragment of the FMG-1 monoclonal antibody) of the probes fail to exhibit enough binding to yield a useful signal.

The principal monoclonal antibody (FMG-1) that has been utilized for the redistribution studies on *C. reinhardtii* strain pf-18 vegetative cells (Bloodgood et al., 1986) exhibits two other interesting properties: (1) This monoclonal antibody does not appear to be able to bind to motile flagella; the flagella of wild-type swimming cells will, after fixation or removal from the cell, label with the antibody. This property has limited the study of the redistribution of the antibody to paralyzed flagella strains. (2) The other property of the antibody, alluded to earlier, is that it inhibits flagellar surface motility (expressed as polystyrene microsphere movements) although it does not affect microsphere binding to the flagellar surface. Thus, the same monoclonal antibody that inhibits rapid, local microsphere movements can be used to visualize the slower, bulk redistribution phenomenon. This suggests that the two dynamic flagellar surface properties discussed in Sections III.B.1 and III.C.2 are fundamentally different in their mechanisms. Both processes are totally inhibited at 4°C and require that the flagellum be attached to a live cell. The rate of glycoprotein redistribution as visualized with the antibody or the lectin is strongly temperature dependent (Bloodgood et al., 1986), whereas the rate of microsphere movement is independent of temperature. Only one of the two processes (flagellar surface motility) appears to be calcium dependent (Bloodgood et al., 1979, 1986). The current working hypothesis is that they are based on two different force-transducing mechanisms associated with the *C. reinhardtii* flagellar surface, although both of the energy-transducing systems may involve an association with the cytoplasmic domain of the 350-kd population of flagellar membrane glycoproteins.

2. Gametes and Fertilization

Mating in *Chlamydomonas* involves an initial random interaction of the flagellar surfaces of + and − gametes by means of highly specific agglutinin molecules (Adair, 1985; van den Ende, 1985). These flagellar contact sites migrate until proper flagellar alignment is achieved (Mesland, 1976), at which time the flagellar tips are in register and a "tip-locking" event occurs. The tip-locking event is followed by the propagation of a signal from the flagella to the cell body initiating cell-wall release and mating structure activation, which are necessary prerequisites for cell fusion (Snell, 1985). The flagellar realignment process can be visualized when gametic cells interact with polystyrene microspheres (Hoffman and Goodenough, 1980) and when gametic cells of one mating type are incubated with flagellar membrane preparations (isoagglutinin or gamone) obtained from gametes of the opposite mating type (Mesland and van den Ende, 1979). It could be that this flagellar realignment reflects the same flagellar surface motor utilized for microsphere movement and gliding motility (Section III.B), or it could reflect a slower, more directed migration of aggregates of specific membrane proteins (such as the agglutinin molecules) within the plane of the flagellar membrane. Hoffman and Goodenough (1980) observed that the movement of flagellar contact sites in mating gametes was slower than the movement of polystyrene microspheres along the gametic flagellar surface. Forest et al. (1978, p. 82) predicted that "gametic flagella possess a vectorial concentration mechanism which, when activated by sexual or antiserum agglutination, brings dispersed signaling components in the flagellar membrane toward the flagellar tip, the signaling reaction taking place when sufficient density of components builds up at the tips." Mating in *C. reinhardtii* is accompanied by the accumulation of dense material on the inside of the flagellar membrane at the distal tips of the flagella (Mesland et al., 1980). Goodenough and Jurivich (1978) observed the accumulation of dense material on the outside of the flagellar membrane at the flagellar tips of gametic cells incubated with rabbit antisera obtained by the injection of rabbits with either vegetative or gametic flagella of either mating type; the authors interpreted these aggregates as antibody that had bound to and been redistributed along the flagellar surface. None of the antisera used was gamete specific; the authors state that the major protein species recognized by each of the four antisera was the HMW flagellar membrane glycoproteins. However, these authors claimed that the accumulation of antibodies only occurred on gametic flagella. This is surprising in light of the results of Bloodgood et al. (1986), discussed above in the Section III.C.1, that monoclonal antibodies recognizing the HMW flagellar membrane glycoproteins of *C. reinhardtii* clearly redistributed on flagella of vegetative cells.

Recently, Drs. Herman van den Ende, Wieger Homan, and their colleagues at the University of Amsterdam have observed the redistribution of components in the flagellar membrane of gametic cells of *C. eugametos.* Figure 9 shows an unmated mating type – gamete that has redistributed a monoclonal antibody recognizing the mating type – agglutinin molecule to the tips of the flagella. More significantly, Dr. van den Ende's laboratory has observed that initiation of mating by the mixing of mating type + and

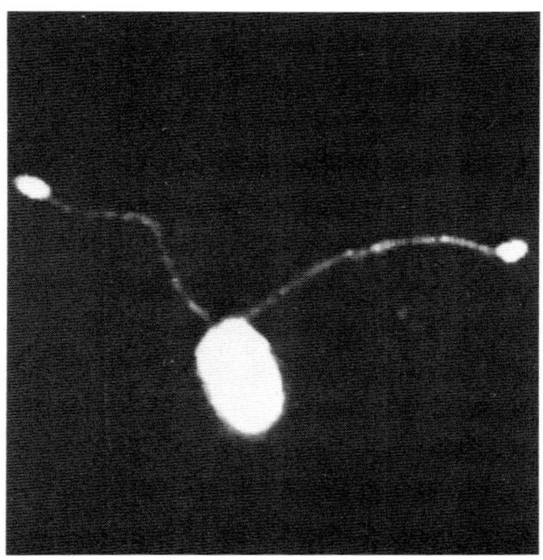

Figure 9. *Chlamydomonas eugametos* gametic cell exhibiting redistribution of a monoclonal antibody recognizing the mating type – agglutinin. The antibody was initially distributed uniformly along the flagellar surface of this unmated gametic cell. (The experiment was performed by Dr. W. Homan and Dr. H. van den Ende at the University of Amsterdam. The photograph was taken with a "confocal scanning light microscope" by Dr. E. van Spronsen, Department of Electron Microscopy and Molecular Cytology, University of Amsterdam.) ×1,200.

mating type – gametic cells induces the redistribution of flagellar membrane proteins (including the mating type – sexual agglutinin glycoprotein) that were initially distributed uniformly along the flagellar surface. These components accumulate at the tips of the gametic flagella; later in the mating process, after flagellar de-adhesion has occurred, these tip aggregates disappear from the flagella. In these studies, the distribution of the flagellar membrane components is visualized after fixation using monoclonal antibodies; therefore, the redistribution of the flagellar proteins recognized by these antibodies cannot be due to the interaction of the antibody with the flagellar surface. This mating-associated redistribution could be a reflection of one of a number of the early events in mating: flagellar alignment (migration of the flagellar contact sites), flagellar tip activation, or the mating induced turnover of flagellar membrane components discussed in Section III.A.2. Flagellar tip activation in *C. reinhardtii* (Mesland et al., 1980) has been shown to be associated with the accumulation of electron-dense material on the inside of the flagellar membrane at the distal end of the flagellum; this material could reflect the tip aggregation of particular flagellar membrane proteins such as the agglutinin molecules.

3. Similar Phenomena in Other Cell Types

In many ways, the redistribution of antibodies and lectins along the flagellar surface of vegetative and gametic *Chlamydomonas* resembles the redistribution of these same ligands that has been reported to occur on the surfaces of lymphocytes and a number of other mammalian and nonmammalian cell types, an energy-dependent process that has been referred to as "capping" (dePetris, 1977). Many, if not most, cases of capping have been shown to be associated with cytoskeletal components: actin, myosin, spectrin, and α-actinin (Bourguignon and Bourguignon, 1984). In general, these particular cytoskeletal components are thought to be absent from cilia and flagella. *Chlamydomonas reinhardtii* flagella have been reported to contain a nonfilamentous, actin-like protein that appears to be associated with the inner dynein arms (Piperno and Luck, 1979; Detmers et al., 1985). Despite the presence of this actin-like component, it is likely that a different set of cytoskeletal components and a different energy-transducing molecule will be found to be responsible for the redistribution phenomena observed on the *Chlamydomonas* flagellar surface.

Antibodies to the i antigen associated with the *Paramecium* ciliary membrane appear to redistribute to the tips of the cilia (Barnett and Steers, 1984); this response does not occur with Fab′ fragments of the antibody, suggesting that the cross-linking of that population of membrane proteins by the antibodies is inducing the redistribution. Using monoclonal antibodies that label only restricted domains on the guinea pig sperm surface,

Myles and Primakoff (1984) have shown that the capacitation of guinea pig sperm results in redistribution of the cell-surface-binding sites for a monoclonal antibody designated PT-1. Before capacitation, this antibody labeled only the posterior tail region; after capacitation, the labeling shifted to the anterior tail region of the sperm. These authors performed some clever experiments to demonstrate that the actual membrane antigen migrates from one region of the sperm-cell surface to another adjacent region. Gonnella and Neutra (1985) have observed the redistribution of cationized ferritin along the surface of microvilli on intestinal epithelial cells. The cationized ferritin initially labeled the length of the microvilli uniformly; subsequently the label accumulated into clumps at the base and tip of each microvillus. The label at the tips of the microvilli was shed into the media in association with plasma membrane blebs.

IV. SUMMARY AND CONCLUSIONS

Cilia and flagella possess a number of distinct advantages for the study of plasma membrane dynamics. Ciliated and flagellated cells provide the opportunity to gain ready access to a defined subset of the cell's general plasma membrane and to the cytoskeletal machinery that underlies that plasma membrane domain. Ciliary and flagellar membranes exhibit a number of interesting dynamic phenomena. More is currently known about the *Chlamydomonas* flagellum than about any other source of cilia and flagella; the same can be said about the *Chlamydomonas* flagellar surface. Much of this stems from the experimental advantages of this organism. Not the least of these is that *Chlamydomonas* provides a useful genetic system that has already been well exploited to study flagellar function. In *Chlamydomonas*, the flagella of gametic cells are of critical importance for reproduction, and mating in this organism involves highly specific membrane–membrane adhesion and membrane signaling phenomena associated with the flagellar membrane. In part, at least, due to their accessibility, much more is currently known about the flagellar membrane surface-exposed glycoproteins of *Chlamydomonas* vegetative and gametic cells than about the underlying (literally and figuratively) mechanisms that can move these glycoproteins within the plane of the membrane. The current state of our knowledge about the *C. reinhardtii* flagellar membrane suggests that the 350-kd class of glycoproteins exhibits a surface-exposed substrate-binding domain which also possesses binding sites for the lectin concanavalin A and certain anticarbohydrate monoclonal antibodies. These same molecules presumably cross the lipid bilayer and possess a cytoplasmic domain that interacts with an axoneme-anchored, energy-transducing system that is capable of moving these glycoproteins laterally in the plane of the flagellar membrane (Figure

5). Because of their high degree of specificity, monoclonal antibodies are proving to be useful probes for studying the organization and dynamics of gametic and vegetative flagellar surfaces and hold promise for use in selecting cell lines defective in defined cell-surface-exposed glycoproteins. An understanding of the physiology of the motors that drive flagellar surface dynamic events will likely come from studies of mutant cell strains with defects in their energy-transducing systems.

ACKNOWLEDGMENTS

The author gratefully acknowledges the following individuals for permission to reproduce their photographs in this review: Drs. Irene Manton, Greg May, Lester I. Binder, Wieger Homan, Herman van den Ende, and Mr. Frederick Reinhart.

Dr. Jonathan Jarvik of Carnegie-Mellon University and the National Science Foundation supported *Chlamydomonas* Genetics Center at Duke University provided strains of *Chlamydomonas* utilized in the author's research described in this review.

The author's research was supported by National Institutes of Health Research Grant GM-28766 and National Science Foundation Research Grant DCB-8502980. Excellent technical assistance was provided by Ms. Nancy Salomonsky. Dr. Charles D. Little generously provided access to his fluorescence microscope.

REFERENCES

Adair, W. S. (1985) J. Cell Sci. (Suppl.) *2:* 233–260.
Barber, V. C. (1974) in *Cilia and Flagella* (M. A. Sleigh, ed.), pp. 403–433. Academic Press, New York.
Bardele, C. F. (1981) Biosystems *14:* 403–421.
Barnett, A., and Steers, E., Jr. (1984) J. Cell Sci. *65:* 153–162.
Bergman, K., Goodenough, U. W., Goodenough, D. A., Jawitz, J., and Martin, H. (1975) J. Cell Biol. *67:* 606–622.
Bessen, M., Fay, R. B., and Witman, G. B. (1980) J. Cell Biol. *86:* 446–455.
Bloodgood, R. A. (1977) J. Cell Biol. *75:* 983–989.
Bloodgood, R. A. (1978) Cell Biol. Int. Rep. *2:* 171–176.
Bloodgood, R. A. (1980) J. Exp. Zool. *213:* 293–295.
Bloodgood, R. A. (1981) Protoplasma *106:* 183–192.
Bloodgood, R. A. (1982) Symp. Soc. Exp. Biol. *35:* 353–380.
Bloodgood, R. A. (1984) Exp. Cell Res. *150:* 488–493.
Bloodgood, R. A., and Levin, E. N. (1983) J. Cell Biol. *97:* 397–404.
Bloodgood, R. A., and May, G. S. (1982) J. Cell Biol. *93:* 88–96.
Bloodgood, R. A., and Workman, L. J. (1984) Cell Motil. *4:* 77–87.
Bloodgood, R. A., Leffler, E. M., and Bojczuk, A. T. (1979) J. Cell Biol. *82:* 664–674.
Bloodgood, R. A., Woodward, M. P., and Salomonsky, N. L. (1986) J. Cell Biol. *102:* 1797–1812.
Bouck, G. B. (1972) Adv. Cell Mol. Biol. *2:* 237–271.
Bouck, G. B., Rogalski, A. A., and Valaitis, A. (1978) J. Cell Biol. *77:* 805–826.

Bourguignon, L. Y. W., and Bourguignon, G. J. (1984) Int. Rev. Cytol. *87:* 195-224.

Bowser, S. S., and Bloodgood, R. A. (1984) Cell Motil. *4:* 305-314.

Bowser, S. S., and Rieder, C. L. (1985) Can. J. Biochem. Cell Biol. *63:* 608-620.

Bowser, S. S., Israel, H. A., McGee-Russell, S. M., and Rieder, C. L. (1984) Cell Biol. Int. Rep. *8:* 1051-1063.

Bowser, S. S., McGee-Russell, S. M., and Rieder, C. L. (1985) Tissue & Cell *16:* 823-839.

Brooker, B. E. (1971) Protoplasma *73:* 191-202.

Ciccolello, R. F., and Gibor, A. (1978) J. Exp. Zool. *206:* 303-306.

Cooper, J. B., Adair, W. S., Mecham, R. P., Heuser, J. E., and Goodenough, U. W. (1983) Proc. Natl. Acad. Sci. USA *80:* 5898-5901.

Crouau, Y. (1980) Biol. Cellulaire *39:* 349-352.

Crouau, Y. (1983) J. Submicrosc. Cytol. *15:* 295-299.

Davies, D. R., and Plaskitt, A. (1971) Genet. Res. *17:* 33-43.

Dentler, W. L. (1981) Int. Rev. Cytol. *72:* 1-47.

Dentler, W. L., Pratt, M. M., and Stephens, R. E. (1980) J. Cell Biol. *84:* 381-403.

dePetris, S. (1977) in *Dynamic Aspects of Cell Surface Organization* (G. Poste and G. L. Nicolson, eds.), pp. 643-728. Elsevier/North-Holland, Amsterdam and New York.

Detmers, P. A., and Condeelis, J. S. (1986) Exp. Cell Res. *163:* 317-326.

Detmers, P. A., Goodenough, U. W., and Condeelis, J. (1983) J. Cell Biol. *97:* 522-532.

Detmers, P. A., Carboni, J. M., and Condeelis, J. (1985) Cell Motil. *5:* 415-430.

Dunlap, K. (1977) J. Physiol. (Lond.) *271:* 119-133.

Forest, C. L., Goodenough, D. A., and Goodenough, U. W. (1978) J. Cell Biol. *79:* 74-84.

Gealt, M. A., Adler, J. H., and Nes, W. R. (1981) Lipids *16:* 133-136.

Gibbons, I. R. (1981) J. Cell Biol. *91:* 107S-124S.

Gitelman, S. E., and Witman, G. B. (1980) J. Cell Biol. *98:* 764-770.

Gonnella, P., and Neutra, M. R. (1985) Anat. Rec. *213:* 520-528.

Goodenough, U. W., and Jurivich, D. (1978) J. Cell Biol. *79:* 680-693.

Goodenough, U. W., Adair, W. S., Calgor, E., Forest, C. L., Hoffman, J. L., Mesland, D. A. M., and Speth, S. (1980) in *Membrane-Membrane Interactions* (N. B. Gilula, ed.), pp. 131-152. Raven Press, New York.

Gumbiner, B., and Louvard, D. (1985) Trends Biochem. Sci. *10:* 435-438.

Hoffman, J. L., and Goodenough, U. W. (1980) J. Cell Biol. *86:* 656-665.

Hyams, J. S., and Borisy, G. G. (1978) J. Cell Sci. *33:* 235-253.

Jahn, T. L., Landman, M. D., and Fonseca, J. R. (1964) J. Protozool. *3:* 291-296.

Jarvik, J. W., and Rosenbaum, J. L. (1980) J. Cell Biol. *85:* 258-272.

Jarvik, J. W., Reinhart, F. D., Kuchka, M. R., and Adler, S. A. (1984) J. Protozool. *31:* 199-204.

Kanno, F., and Ishii, K. (1979) Bull. Fac. Liberal Arts, Hosei Univ. *31:* 1-8.

Kitamura, A., and Hiwatashi, K. (1984) Zool. Sci. *1:* 161-168.

Kung, C., and Saimi, Y. (1982) Annu. Rev. Physiol. *44:* 519-534.

Lee, A. G. (1978) Int. Rev. Biochem. *19:* 221-270.

Levin, R. M., and Weiss, B. (1979) J. Pharmacol. Exptl. Ther. *208:* 454-459.

Lewin, R. A. (1952) Biol. Bull. *103:* 74-79.

Lewin, R. A. (1954) J. Gen. Microbiol. *11:* 358-368.

Lewin, R. A. (1982) Experientia *38:* 348-349.

Low, P. S., Lloyd, D. H., Stein, T. M., and Rogers, J. A. (1979) J. Biol. Chem. *254:* 4119-4125.

Machemer, H., and Ogura, A. (1979) J. Physiol (Lond.) *296:* 49-60.

McLean, R. J., Laurendi, C. J., and Brown, R. M., Jr. (1974) Proc. Natl. Acad. Sci. USA *71:* 2610-2613.

Manton, I. (1965) Advan. Botan. Res. *2:* 1-34.

Marchese-Ragona, S. P., and Johnson, K. A. (1985) J. Cell Biol. *101:* 278a.

Markey, D. R., and Bouck, G. B. (1977) J. Ultrastruct. Res. *59:* 173-177.

Melkonian, M. (1984) in *Systematics of the Green Algae* (D. E. G. Irvine, ed.), pp. 73–120. Academic Press, New York.
Mesland, D. A. M. (1976) Arch. Microbiol. *109:* 31–35.
Mesland, D. A. M., and van den Ende, H. (1979) Protoplasma *98:* 115–129.
Mesland, D. A. M., Hoffman, J. L., Caligor, E., and Goodenough, U. W. (1980) J. Cell Biol. *84:* 599–617.
Mitchell, D. R., and Rosenbaum, J. L. (1985a) J. Cell Biol. *100:* 1228–1234.
Mitchell, D. R., and Rosenbaum, J. L. (1985b) J. Cell Biol. *101:* 278a.
Molyneaux, D. H. (1969) Parasitology *59:* 55–66.
Monk, B. C., Adair, W. S., Cohen, R. A., and Goodenough, U. W. (1983) Planta *158:* 517–533.
Moran, D. T., Varela, F. J., and Rowley, J. Carter, III (1977) Proc. Natl. Acad. Sci. USA *74:* 793–797.
Myles, D. G., and Primakoff, P. (1984) J. Cell Biol. *99:* 1634–1641.
Pijst, H. L. A. (1985) Ph.D. dissertation, University of Amsterdam, Chapt. 4.
Pijst, H. L. A., Zilver, R. J., Musgrave, A., and van den Ende, H. (1983) Planta *158:* 403–409.
Pijst, H. L. A., Ossendorp, F. A., van Egmond, P., Kamps, A. M. I. E., Musgrave, A., and van den Ende, H. (1984) Planta *160:* 529–535.
Piperno, G., and Luck, D. J. L. (1979) J. Biol. Chem. *254:* 2187–2190.
Rhein, L. D., and Cagan, R. H. (1980) Proc. Natl. Acad. Sci. USA *77:* 4412–4416.
Reinhart, F. D., and Bloodgood, R. A. (1984) J. Cell Biol. *99:* 178a.
Remillard, S. P., and Witman, G. B. (1982) J. Cell Biol. *93:* 615–631.
Ringo, D. L. (1967) J. Cell Biol. *33:* 543–571.
Roberts, K., Grief, C., Hills, G. J., and Shaw, P. J. (1985) J. Cell Sci. (Suppl.) *2:* 105–127.
Rogalski, A. A., and Bouck, G. B. (1980) J. Cell Biol. *86:* 424–435.
Rohlich, P. (1975) Cell Tissue Res. *161:* 421–430.
Rosenbaum, J. L., and Child, F. M. (1967) J. Cell Biol. *34:* 345–364.
Saito, T., and Matsuda, Y. (1984a) Exp. Cell Res. *152:* 322–330.
Saito, T., and Matsuda, Y. (1984b) Arch. Microbiol. *139:* 95–99.
Saito, T., Tsubo, Y., and Matsuda, Y. (1985) Arch. Microbiol. *142:* 207–210.
Satir, B., Sale, W. S., and Satir, P. (1976) Exp. Cell Res. *97:* 83–91.
Seeman, P. (1972) Pharmacol. Rev. *24:* 583–654.
Snell, W. J. (1976) J. Cell Biol. *68:* 48–69.
Snell, W. J. (1981) J. Supramol. Struct. Cell Biochem. *16:* 371–376.
Snell, W. J. (1983) Cell Motil. *3:* 273–280.
Snell, W. J. (1985) Annu. Rev. Plant Physiol. *36:* 287–315.
Snell, W. J., and Moore, W. S. (1980) J. Cell Biol. *84:* 203–210.
Snell, W. J., and Roseman, S. (1979) J. Biol. Chem. *254:* 10820–10829.
Snell, W. J., Buchanan, M., and Clausell, A. (1982) J. Cell Biol. *94:* 607–612.
Snell, W. J., Clausell, A., and Moore, W. S. (1983) J. Cell Biol. *96:* 589–597.
Solter, K. M., and Gibor, A. (1978) Exp. Cell Res. *115:* 175–181.
Suzaki, T., and Shigenaka, Y. (1982) in *Biological Functions of Microtubules and Related Structures* (H. Sakai, H. Mohri, and G. G. Borisy, eds.), pp. 91–103. Academic Press, New York.
Suzaki, T., Shigenaka, Y., Watanabe, S., and Toyohara, A. (1980) J. Cell Sci. *42:* 61–79.
Troyer, D. (1975) Nature (Lond.) *254:* 696–698.
Tuomanen, E. I., and Hendley, J. O. (1983) J. Infect. Diseases *148:* 125–130.
Tyler, S. (1973) Acta Zool. *54:* 139–151.
van den Ende, H. (1985) Adv. Microbial Physiol. *26:* 89–123.
Vickerman, K. (1973) J. Protozool. *20:* 394–404.
Watanabe, T. (1977) J. Protozool. *24:* 426–429.
Watanabe, T., and Flavin, M. (1976) J. Biol. Chem. *251:* 182–192.

Weiss, R. L., Goodenough, D. A., and Goodenough, U. W. (1977) J. Cell Biol. *72:* 133–143.
Wiese, L. and Hayward, P. C. (1972) Am. J. Bot. *59:* 530–536.
Wiese, L., Hardcastle, A., and Stewart, M. (1984) Gamete Res. *9:* 441–449.
Witman, G. B. (1975) Ann. NY Acad. Sci. *253:* 178–191.
Witman, G. B., Carlson, K., Berliner, J., and Rosenbaum, J. L. (1972) J. Cell Biol. *54:* 507–539.

A STRUCTURAL ANALYSIS OF ONE PHOTOSYNTHETIC MEMBRANE: *RHODOPSEUDOMONAS VIRIDIS*

Kenneth R. Miller

I. INTRODUCTION

Biological membranes are involved with a number of fundamental processes in living organisms. None of these is more basic than the roles that membranes play as energy-conversion devices. While nearly all cellular membranes are capable of carrying out transport functions which are energy related, the specialized membranes of mitochondria and chloroplasts are of particular interest. These membranes contain a series of electron transport components which produce strong electrochemical gradients, and these gradients provide the driving force for the phosphorylation of adenosine diphosphate to produce adenosine triphosphate (ATP).

Here I will describe results from my laboratory that have focused on the organization of photosynthetic membrane components. The major

Advances in Cell Biology, Vol. 1, pages 131–156.

ISBN: 0-89232-792-8

biochemical and energetic events which take place in photosynthetic membranes are the light reactions of photosynthesis. In addition to the electron transport and ATP-synthesizing functions which these membrane have in common with inner mitochondrial membranes, photosynthetic membranes must also contain systems for harvesting and utilizing sunlight. In such membranes, light energy is used to promote a primary charge separation in pigment molecules within a reaction center complex. The separation of charge provides the driving force for subsequent electron transport and the ultimate conversion of solar energy into chemical energy.

II. THE ORGANIZATION OF PHOTOSYNTHETIC MEMBRANES

The general features of photosynthetic membrane organization have been reviewed elsewhere (Miller, 1985) and will not be covered in detail here, except to provide a context for recent work. The photosynthetic membranes of higher plants are complex structures that display an intricate internal organization, are composed of more than 60 membrane-associated polypeptides, and contain at least two distinct types of photosynthetic reaction centers. While work on such membranes continues at a rapid rate, the very complexity of the system has presented problems for many types of studies. Therefore, a number of laboratories have focused at least part of their attention on the development of experimental systems in which the basic features of photosynthetic membrane organization can be investigated more intensively.

The photosynthetic bacteria have been widely used as such systems, and they present a number of advantages over higher plants for certain types of studies. Photosynthetic bacteria differ from high plants in a number of ways, including the possession of a single photosystem rather than two. These organisms also contain photosynthetic reaction centers that can be more easily isolated and characterized than higher plant systems, and a considerable body of knowledge has been developed on their structure, organization, and the photosynthetic processes which they carry out (Clayton and Sistrom, 1978).

Our laboratory has worked on several photosynthetic bacteria, but one species in particular has presented such advantages for study that it has become the central point of our research effort on prokaryotic photosynthetic membranes: *Rhodopseudomonas viridis*, a purple, nonsulfur photosynthetic bacterium. It is an obligate anaerobe, and uses bacteriochlorophyll b as its primary photopigment. Attention was first attracted to this organism as a result of a report by Giesbrecht and Drews (1966) regarding the structure of its internal photosynthetic membranes. The apparent hexagonal symmetry displayed in electron micrographs of these membranes implied that image analysis techniques could be used to obtain detailed structural information about this organism. Because the photosynthetic membranes in higher plants do not display such symmetry, it seemed that *Rh. viridis* might present considerable advantages for the study of photosynthetic membrane organization.

III. ORGANIZATION OF THE PHOTOSYNTHETIC MEMBRANE

Figure 1 illustrates the general structure of *Rh. viridis* and its photosynthetic membranes, which are visible within the cell as a tightly packed stack. The cells can be broken easily and photosynthetic membranes isolated via differential centrifugation (Jacob and Miller, 1983). Electron micrographs of negative-stained material show the regular pattern of organization which makes this organism so interesting (Figure 2). Fourier analysis techniques which have been well described elsewhere (DeRosier and Klug, 1966) make

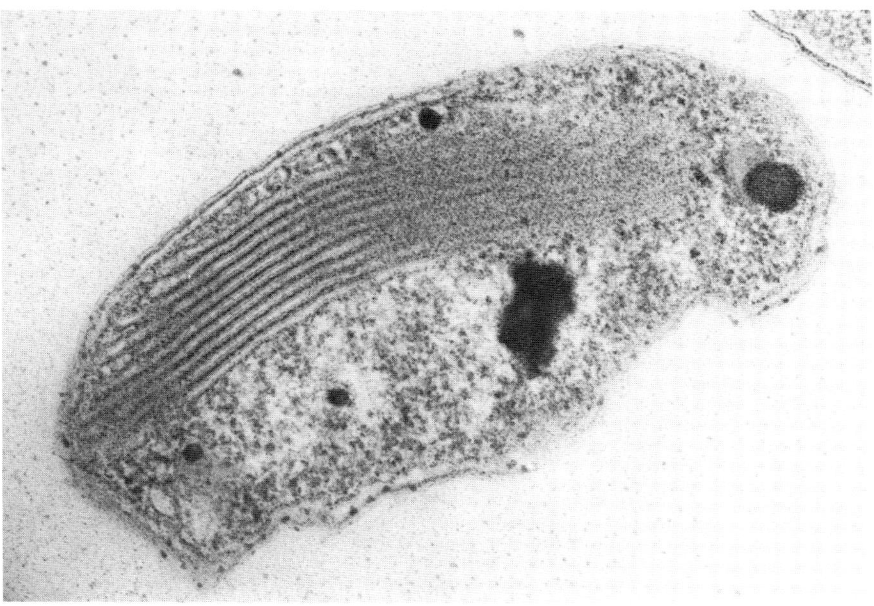

Figure 1. The photosynthetic membranes of *Rh. viridis* are visible in this thin section of a cell as a tight stack of membranes. These membranes can be isolated from disrupted cells and used for study. ×58,000.

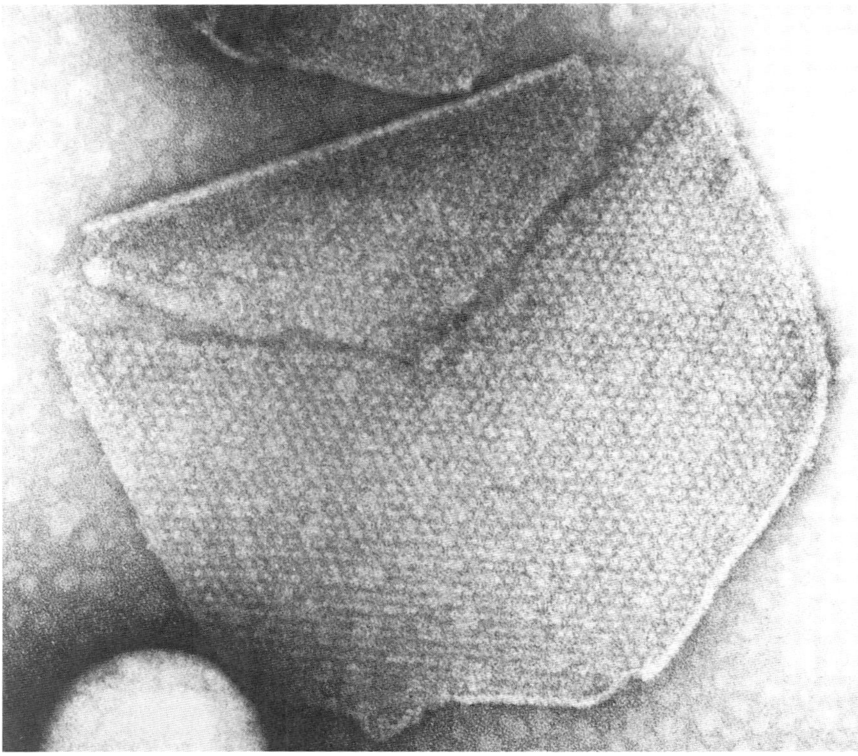

Figure 2. Isolated photosynthetic membranes can be examined by negative staining. In this sample, contrasted with 2% uranyl acetate, the regular substructure of the membrane is apparent. The repeating features of this membrane make it an ideal subject for image enhancement by Fourier techniques. ×156,000.

it possible to enhance the detail observed in such images and to prepare two-dimensional maps of the distribution of stain-excluding mass within the membrane subunits (Miller, 1979). Although it is possible to image the photosynthetic membranes of *Rh. viridis* in a number of ways, including thin-sectioning and freeze-etching, negative staining has been the most fruitful, partly because negatively stained images can be used directly in Fourier analysis studies.

The individual subunits, which are arranged at a center-to-center spacing of approximately 12.0 nm, display a large central stain-excluding mass surrounded by a ring of lesser density (Figure 3). This image is supported by quick-freeze and deep-etch micrographs of the membrane outer surface,

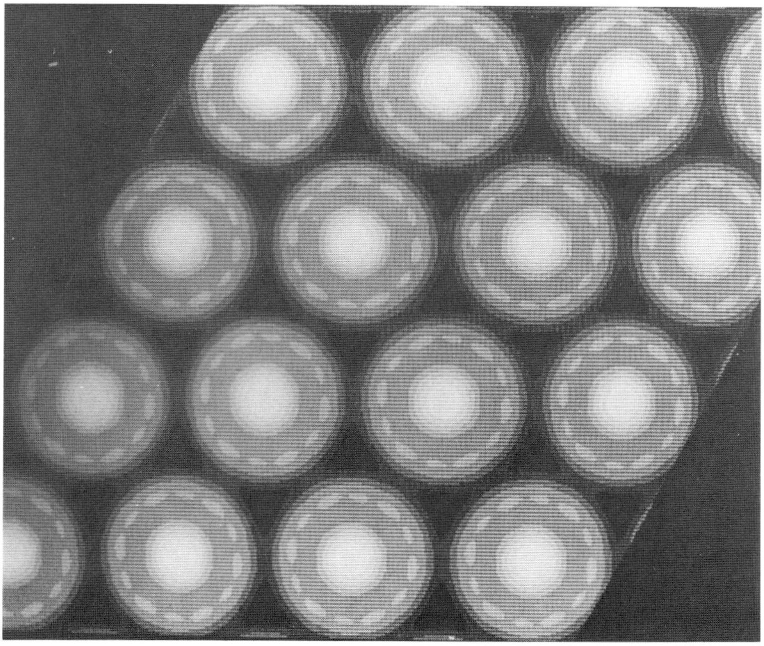

Figure 3. A filtered image of the *Rh. viridis* photosynthetic membrane prepared by Fourier filtering techniques from negatively stained membranes. In this image, the stain-excluding regions are shown in white. Each subunit in the membrane is composed of a large central structure surrounded by a ring of smaller structures. The center-to-center distance between each subunit is approximately 130 Å.

as shown in Figure 4. We have used the Fourier techniques to produce a three-dimensional model of the membrane as well, utilizing a series of electron micrographs that were collected at tilt angles in the microscope ranging from 0° to 61°. The combined Fourier terms from these micrographs were used to form a three-dimensional data set from which the structure of an individual membrane subunit could be determined. Figure 5 shows the result of this reconstruction (Miller, 1982): a structure that confirms the presence of a large membrane-spanning central structure surrounded by a ring of subunits more tightly embedded in the membrane itself.

The images that were used in this study contained four diffraction orders in the Fourier transforms, but only the first three orders were used for the reconstruction. This gives the final image an approximate resolution of 33 Å, sufficient to discern the general size and shape of the subunit, but

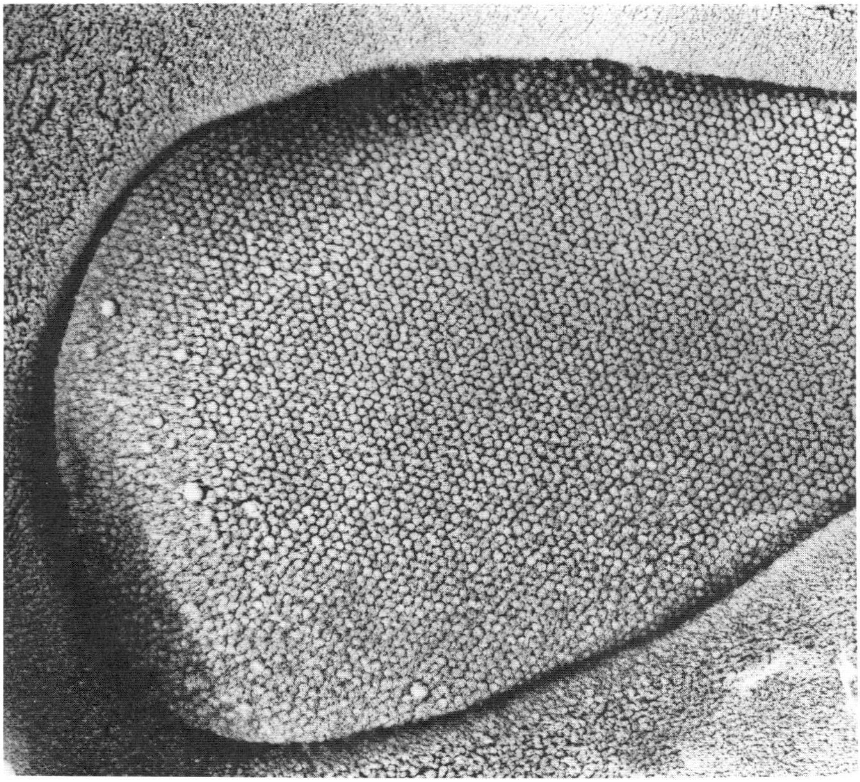

Figure 4. The surface of the *Rh. viridis* photosynthetic membrane can be visualized by quick-freezing and deep-etching. In this image, the individual subunits are clearly visible, and it is apparent that the central structure in each subunit protrudes from the surface of the membrane. ×125,000.

revealing few internal details of the basic structure. A recent paper (Stark et al., 1984) claimed images of the *Rh. viridis* photosynthetic membrane in which meaningful data could be obtained from the sixth order of the Fourier transform. When the Fourier-filtered map of this membrane was compared to a previously published three-dimensional map (Miller, 1982), the authors claimed that significant differences were noted. The major reason for the improved resolution obtained by Stark et al. (1984) appears to be the use of a low-dose unit attached to the electron microscope, which limits the amount of electron radiation absorbed by the sample during image formation. Low-dose experiments in our laboratory have confirmed the basic observations of these workers. Figure 6a shows a central section through

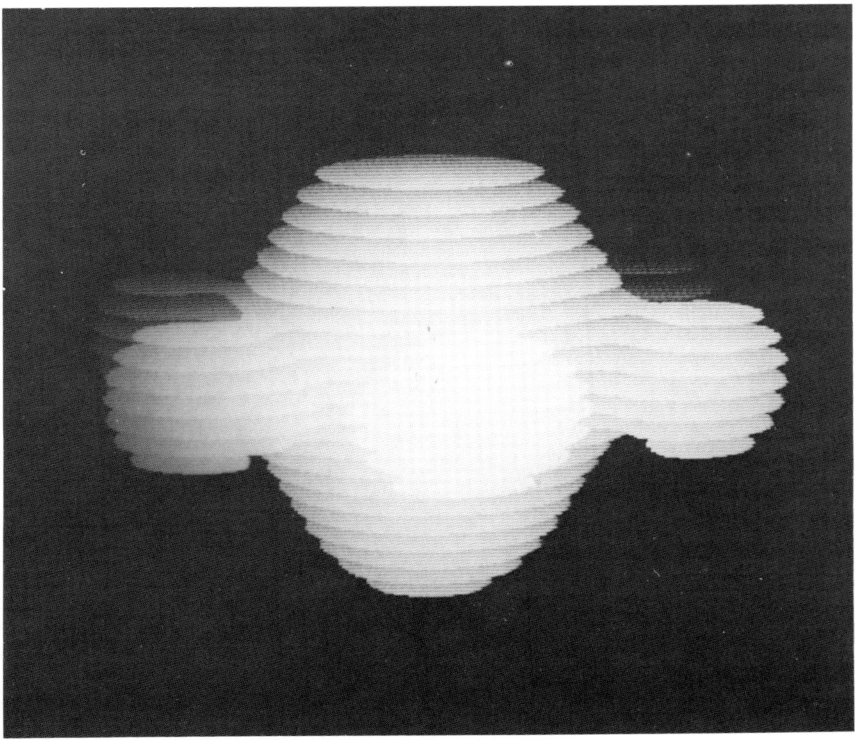

Figure 5. A series of electron micrographs of the *Rh. viridis* photosynthetic membrane, taken at various tilt angles, were used to form a three-dimensional data set from which a complete reconstruction of the membrane subunit could be produced. This computer-generated image shows the outline of such a subunit, with a large central mass surrounded by a ring of smaller structures. The structure is oriented so that the membrane in which it is found lies in the horizontal direction. The low (ca. 33-Å) resolution of the structure does not allow the individual structures in the outer ring to be distinctly separated.

the three-dimensional model shown in Figure 5 (Miller, 1982). The organization of the subunit is dominated by a large central structure, and it is surrounded by six peripheral densities.

Because the 1982 study was done using the first three orders of the Fourier transform, we have recently prepared images in which electron-dose limitation has increased resolution to the sixth order of the Fourier transform, as described by Stark et al. (1984). Following their procedure, we have

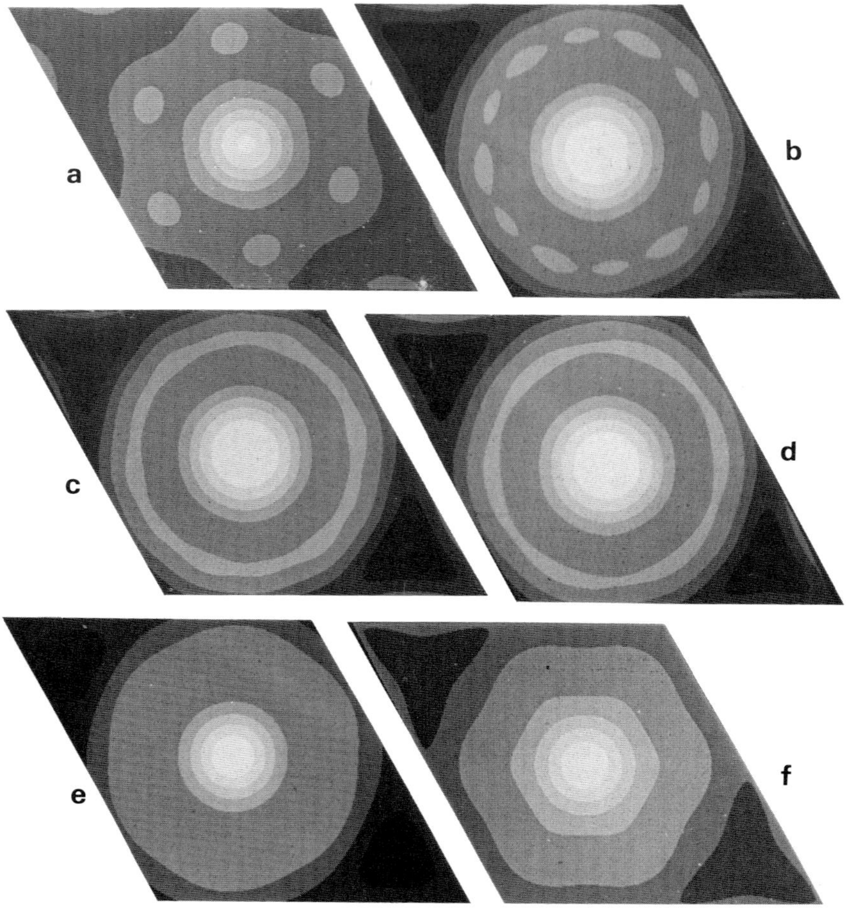

Figure 6. (a) A central section through the three-dimensional map of the *Rh. viridis* photosynthetic membrane (Miller, 1982). Stain-excluding regions are seen as light shades, and heavily stained regions are dark; 2% uranyl acetate was used as the negative stain. (b) A Fourier-filtered image obtained from a two-dimensional projection of the *Rh. viridis* photosynthetic membrane at zero tilt. Transform data significant to six orders were included. The data were first refined to produce an appropriate phase origin and then were averaged for P6 symmetry. A large central structure is visible, surrounded by a ring containing as many as 12 small subunits. (c,d,e) Filtered images obtained from data identical to those for panel, except that Fourier terms beyond the fifth order were discarded in panel c, beyond the fourth order were discarded in panel d, and beyond the third order in panel e.

filtered such images using Fourier terms to the sixth order, averaged the data for sixfold symmetry, and presented the results as shown in Figure 6b. This image is essentially identical to that presented in the recent study (Stark et al., 1984). However, it does differ from a central section through our three-dimensional model of the same membrane. Our recent studies have shown that there are several reasons for these differences.

There are a number of determinants of apparent structure for a three-dimensional model determined from electron microscopy by Fourier techniques. One of these, of course, is the choice of contour level. Several contour levels might be selected from Figure 6a for model-building purposes, and some of these emphasize the apparent six subunits surrounding the central mass more than others. However, the major feature found in the recent high-resolution images (Figure 6b) is the separation of a ring of mass from the large structure in the center of the complex. This feature is clearly related to the resolution at which the data from the Fourier transform are selected. Figures 6c–e show the filtered images that are obtained by restricting the Fourier data used for the reconstruction to five orders, four orders, and three orders, respectively. It is obvious that the existence of this ring is first apparent when fourth-order data are included, that the inclusion of fifth-order data do not change the image substantially, but that sixth-order data do change the detail of the ring, causing smaller subunits to appear. When alternate contour levels for the sixth-order image are chosen, the 12 paired subunits in the ring fuse into six larger ones.

When the zero-tilt data from the *Rh. viridis* three-dimensional reconstruction is used to prepare a filtered image, the result is very similar to that obtained by restricting the current high-resolution data to three orders (compare Figure 6e to Figure 6f). This implies that the appearance of six lobes in the three-dimensional reconstruction is caused by the inclusion of data obtained from tilted images in the Fourier terms. Therefore, the failure to match our three-dimensional model precisely with two-dimensional images (Engelhardt et al., 1983; Stark et al., 1984) is not surprising. It merely points out the fact that additional information may be obtained from a biological sample by analysis of data from tilted specimens.

Interestingly, the filtered images which Start et al. (1984) have published of shadowed preparations of one surface of the *Rh. viridis* membrane do

---

Figure 6—continued
(f) A Fourier reconstruction of the zero-tilt averaged images which were used as a basis for building the three-dimensional data set in our earlier study (Miller, 1982). This untilted image, although it is the basis of the data set from which Figure 1 was taken, is essentially identical to our most recent data (panels b–e) after the Fourier data are restricted to three orders (33 Å).

show six clearly visible peripheral subunits. Similar images of the other surface show twelve subunits. Therefore, one might suggest that the three-dimensional model we have published for the *Rh. viridis* membrane precisely predicts the results of Stark et al. (1984)—namely, the peripheral subunits extending above the membrane surfaces are arranged in ring-like fashion around the central subunit. The data from these recent papers confirm the details of the *Rh. viridis* three-dimensional model (Miller, 1982) in a most elegant way. Engelhardt et al. (1983) have shown that this pattern of membrane structure is a general property of bacteriochlorophyll b-containing organisms, and Stark et al. (1984) have shown that the peripheral subunits surrounding the central projection are indeed exposed at each surface of the membrane. The principal limitation of the 1982 model is indeed the 33-Å (third-order) resolution at which it was prepared. However, the model also contains image details in the axis perpendicular to the plane of the membrane that are not apparent from two-dimensional projected views and have now been confirmed by surface shadowing studies.

Finally, the differences in filtered images of the *Rh. viridis* photosynthetic membrane as a function of higher-order information are substantial (see Figure 6). The higher-order information becomes available largely by limiting electron dose and resultant radiation damage. Better specimen preservation certainly results from such procedures, and resolution is improved. However, a word of caution should attend the results shown in Figure 2. The visualization of 12 subunits in the peripheral ring of material does not occur until the sixth-order data from the Fourier transform are included in the reconstruction. Because these amplitude terms are among the weakest in the transform, we should exercise caution in determining the exact number and position of subunits which surround the central structure. Efforts in our lab are now directed toward preparing a detailed high-resolution map of the membrane under conditions where the preservation of structure will allow for better visualization of components within the membrane.

IV. BIOCHEMICAL STUDIES OF THE MEMBRANE

The *Rh. viridis* photosynthetic membrane contains seven polypeptides, four of which are part of the photosynthetic reaction center (RC), and three low-molecular-weight light-harvesting (LH) polypeptides (Jay et al., 1984). Our interest in probing the topology of the membrane led us to study the susceptibility of the membrane-bound polypeptides to proteolysis. Each of the four reaction center polypeptides were rapidly degraded by both trypsin and pronase, while the LH polypeptides seem to be protected against the enzymes (Jacob and Miller, 1983).

Later studies showed that while proteolysis attacked the RC polypeptides it did not abolish the 830-nm absorbance signal which is characteristic of a functional RC (Jacob and Miller, 1984). Treatment with the nonionic detergent LDAO (LDAO, *N,N*-dimethyldode-cylamine-*N*-oxide), while it dissolved the membrane, also failed to affect the 830-nm signal. However, a combination of proteolysis and detergent treatment quickly destroyed the RC signal. These experiments led us to propose a model for the membrane in which the 830-nm signal was dependent on the interactions of portions of each of the RC polypeptides and in which the reaction center was surrounded by LH polypeptides.

Based on the three-dimensional reconstruction derived from electron microscopy, the most logical positions on the structure to assign for a reaction center would seem to be the large, membrane-spanning central structure. LH components would then be arranged around the outside of the structure in a way that would protect them from proteolytic attack.

V. ISOLATION OF THE PHOTOSYNTHETIC REACTION CENTER

The photosynthetic reaction center of *Rh. viridis* is a membrane protein composed of four polypeptides. Detergent solubilization of photosynthetic membranes combined with hydroxylapatite chromatography yields a single protein containing all four polypeptides and retaining functional photosynthetic activity (Miller and Jacob, 1983). By gradually dialyzing away excess detergent, we were able to form large, flat, crystalline sheets of purified reaction centers that could be visualized in the electron microscope. Figure 7 illustrates the appearance of these sheets, which may contain as many as 10,000 individual subunits and are highly ordered, making detailed Fourier analysis possible. These sheets provide an opportunity to visualize individual photosynthetic reaction centers, and Fourier-filtered images prepared from negatively stained samples showed the reaction center to be a roughly elliptical structure (Figure 8). These sheets can also be visualized in freeze-fractured and deep-etched samples (Figure 9), and some of the features visible in such replicas have been used to infer the placing of individual subunits with the sheet-like crystal (Miller and Jacob, 1985a).

The techniques that were used to produce these sheet-like crystals were similar to methods employed earlier by Michel (1982) to prepared three-dimensional crystals from the *Rh. viridis* reaction center. Because generally it has been impossible to crystallize membrane proteins, the crystallization technique employed by Michel (1983) has been of great interest to other workers who might like to crystallize membrane proteins. These three-dimensional crystals were the basis of high-resolution x-ray diffraction

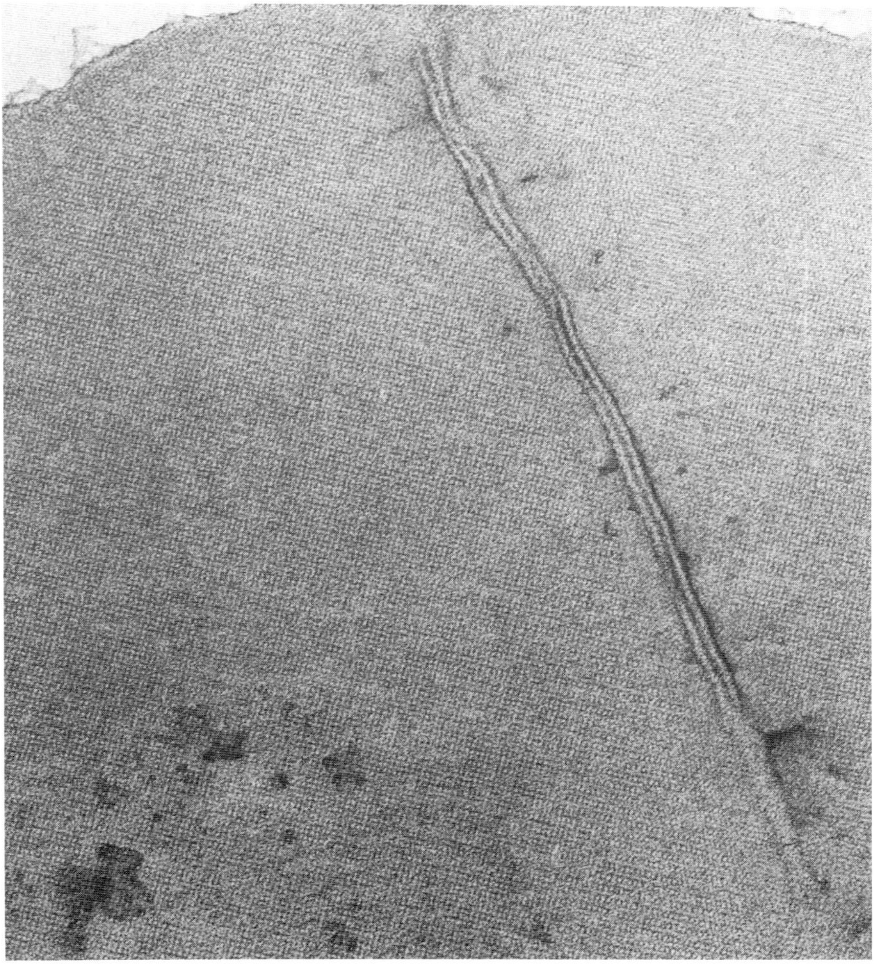

Figure 7. When purified reaction center proteins are dialyzed under appropriate conditions, large, well-ordered crystalline sheets are formed (Miller and Jacob, 1983). Individual reaction centers are visible as small stain-excluding masses in such sheets, and the size and order of the sheets allows them to be used for Fourier image analysis. A fold in the sheet running from top center to the lower right-hand corner provides an edge-on view of the sheet that can be used to provide an estimate of zero-order amplitudes useful in preparaing a three-dimensional reconstruction of the subunits within the sheet. ×95,000.

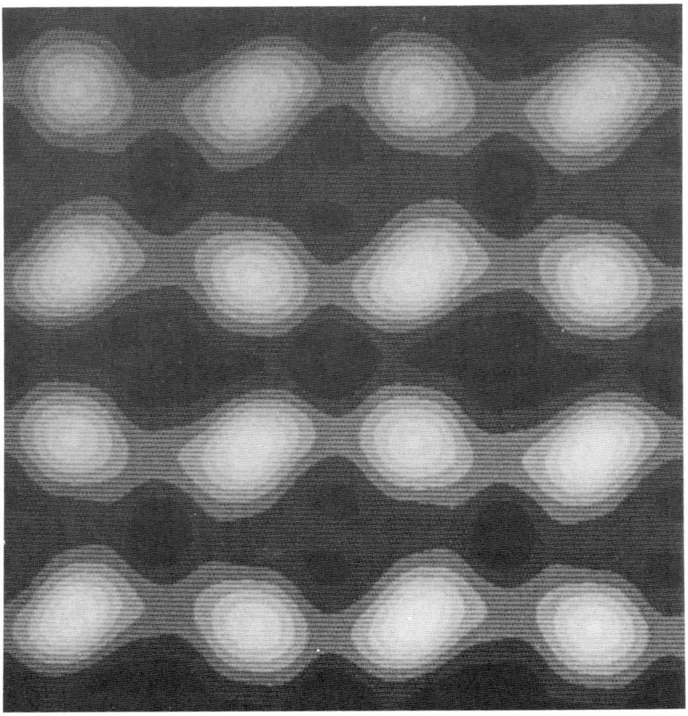

Figure 8. Fourier filtered image of the micrograph shown in Figure 7. The image corresponds to a projection of the entire structure, and four unit cells containing 16 stain-excluding subunits are shown.

patterns, which have now led to a detailed three-dimensional structure for the reaction center protein (Diesenhofer et al., 1984, 1985).

VI. THREE-DIMENSIONAL MODELING OF THE REACTION CENTER

Two-dimensional crystalline sheets of purified *Rh. viridis* reaction centers provide an opportunity to develop a complete three-dimensional map of the RC protein. The data for this analysis was obtained from a set of 37 electron micrographs of RC two-dimensional crystals, contrasted with 2% uranyl acetate. The set of micrographs was collected at tilt angles up to 68°, providing a set of Fourier transforms from which amplitude and phase data could be sampled along 31 different independent lattice lines. An estimate

Figure 9. The two-dimensional crystalline sheets of purified reaction center can also be studied by quick-freezing and deep-etching. The regular pattern that is visible in negative stain can also be seen in this replica, and differences between shadowed and stained images were used to form a preliminary three-dimensional model for the arrangement of subunits within this sheet (Miller and Jacob, 1985). ×196,000.

of the 0,0 lattice line in three dimensions cannot be obtained from tilted specimens, and edge-on views of the RC sheets were used to determine the values for this portion of the three-dimensional transform. Amplitude and phase terms from each micrograph in the data set were combined to form a three-dimensional data set. Images from this data set showed an average phase residual of 24° during the combining process. Representative phase and amplitude data from four of the lattice lines are shown in Figure 10, along with the curves from which values were interpolated to assemble the data used for the reverse three-dimensional transform. The data included terms to the sixth order, indicating a resolution of ~21 Å. Individual micrographs of untilted specimens displayed clear P2 symmetry, and the three-dimensional analysis was carried out with this symmetry applied to the data.

These sheet-like crystals, as shown in Figure 7, consist of individual stain-excluding structures arranged in a rectangular lattice. This contrasts sharply with the organization of the *Rh. viridis* photosynthetic membrane, in which individual subunits are arranged in a hexagonal pattern (with apparent P6 symmetry). The crystallographic unit cell for the reaction center sheets includes four individual stain-excluding units (as shown in Figure 8) and has overall average dimensions of 12.9 × 12.4 nm (Miller and Jacob, 1983). Each of the four subunits within the unit cell has an average cross-section of 5.0 × 3.5 nm. The process of three-dimensional reconstruction by Fourier techniques allows us to determine the arrangement of these four subunits within the unit cell (DeRosier and Klug, 1966). Such a reconstruction gives an accurate image of the actual structure of each subunit, subject to the basic limitations of the technique, namely, the fact that only stain-excluding regions are visible, the specimen is subject to distortion and radiation damage in the electron microscope, and that resolution is limited by the very process of dehydration and staining to ~2 nm in these specimens.

The reconstructed three-dimensional data may be sampled in several ways. Figure 11 shows a contour map illustrating density regions taken through the center of the sheet. Four clearly defined subunits are visible, arranged in a manner that is consistent with the two-dimensional projection of the same structure (Figure 8). Such contours may be taken at intervals to build up a three-dimensional map of the structure. In Figure 12 we have selected a single contour value as the edge of the structure and displayed slices of the structure at regular intervals on a computer graphics device. This technique produces clear images of each of the four subunits within a crystallographic unit cell.

As shown in Figure 12, each subunit has a roughly cylindrical shape, ~5.0 nm in diameter at its thickest point, and extending from one edge of the crystalline sheet to the other. Each subunit is inclined ~20° from the axis normal to the sheet. The length of each subunit in the unit cell is ~85 Å.

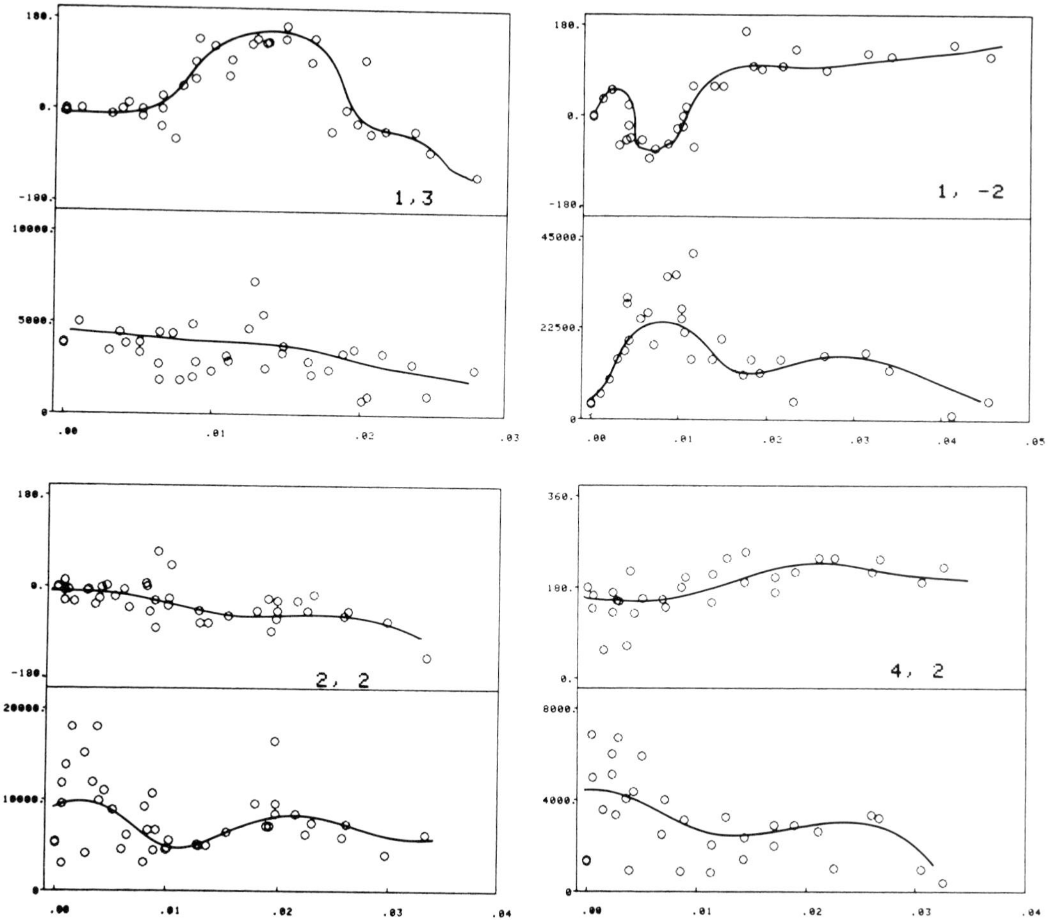

Figure 10. Phase and amplitude plots of four of the lattice lines used for the three-dimensional reconstruction. P2 symmetry was assumed in merging the data from 37 micrographs of negatively stained reaction center sheets.

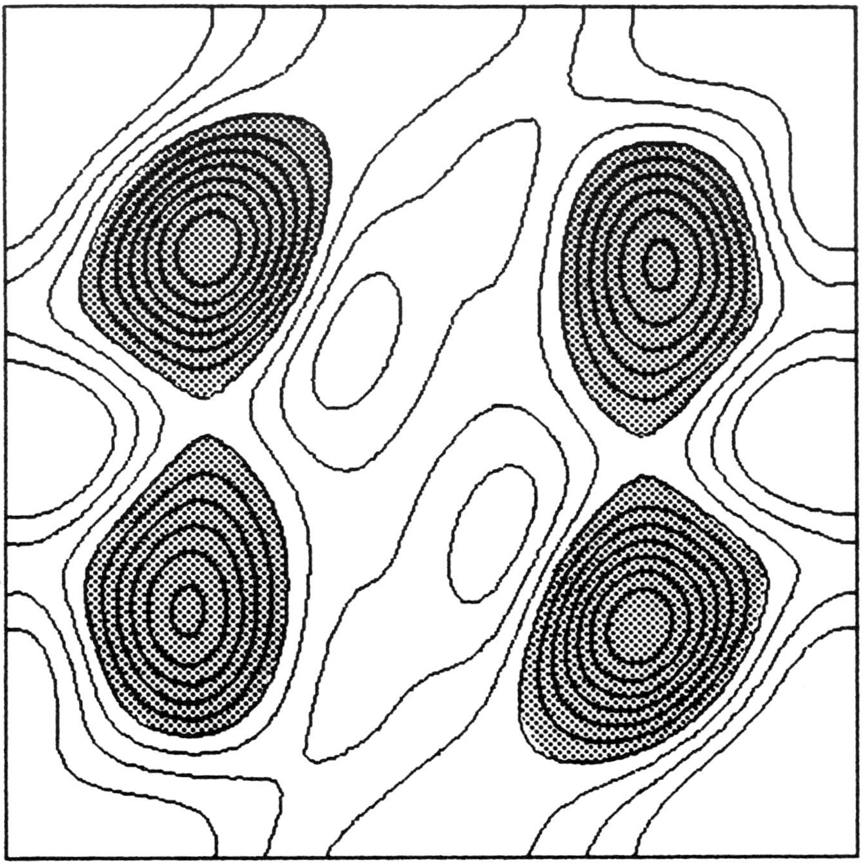

Figure 11. A contour plot through the central region of the three-dimensional map of the reconstructed reaction center crystal. Areas selected at the edge of the structure are shown in shading. At this section, each of the four subunits is positioned near one corner of the unit cell (dimensions: 12.3 nm horizontal; 12.9 nm vertical).

VII. EACH SUBUNIT A COMPLETE REACTION CENTER

The *Rh. viridis* reaction center consists of four polypeptides with molecular weights ranging from 24 to 41 kd as reported by a number of laboratories (Miller and Jacob, 1983; Jay et al., 1984; Pucheu et al., 1976). These four subunits include three polypeptides that are typical of reaction centers described for other purple photosynthetic bacteria and designated H, M,

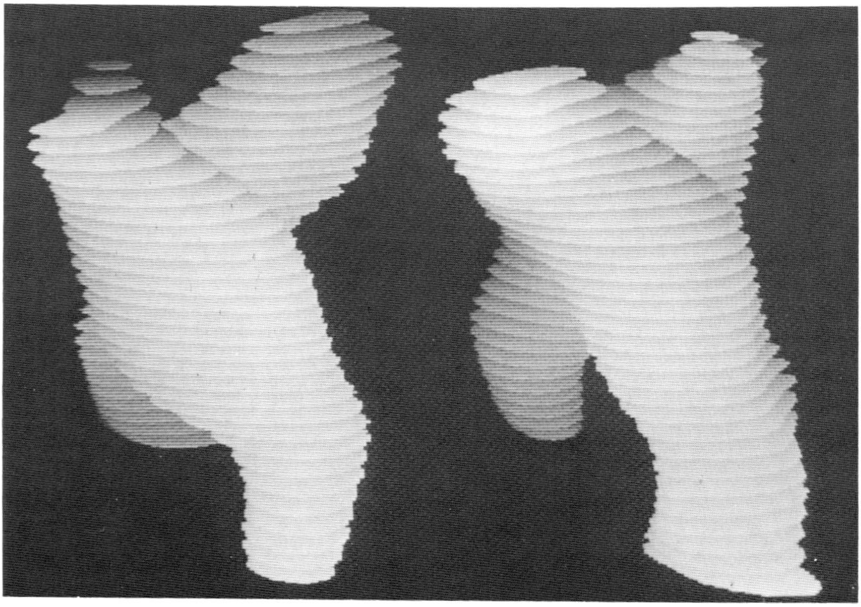

Figure 12. A complete three-dimensional reconstruction of one unit cell of the crystalline sheet. Individual sections through the model have been taken at 0.3-nm intervals. Each of the four reaction centers can be observed. The reaction centers are roughly cylindrical in shape, and are inclined at an angle to the vertical of about 20°. Each reaction center is ~8.5 nm in length. This map was produced by a graphics system developed by the author.

and L (for heavy, medium, and light, respectively) (Parson, 1982). In addition, the *Rh. viridis* photosynthetic membrane contains a fourth polypeptide, which has been identified as a c-type cytochrome containing at least four heme groups (Thornber, 1980). Can each of these four polypeptides be included within a single subunit in the unit cell? Adding together the apparent molecular weights of the polypeptides produces an aggregate of ~125 kd. If one assumes a density of 1.3 $Å^3$/d, then an aggregate molecular volume for the complex of ~162,500 $Å^3$ can be predicted, neglecting the contribution of bound pigments and prosthetic groups to the overall volume. A rectangular solid measuring 5.0 Å × 3.5 Å × 8.5 nm has a volume of 148,750 $Å^3$. While uncertainties abound in estimates made of structures embedded in negative stain, it is clear that the size of a single subunit in the unit cell is comparable to what we might expect from a single photosynthetic reaction center.

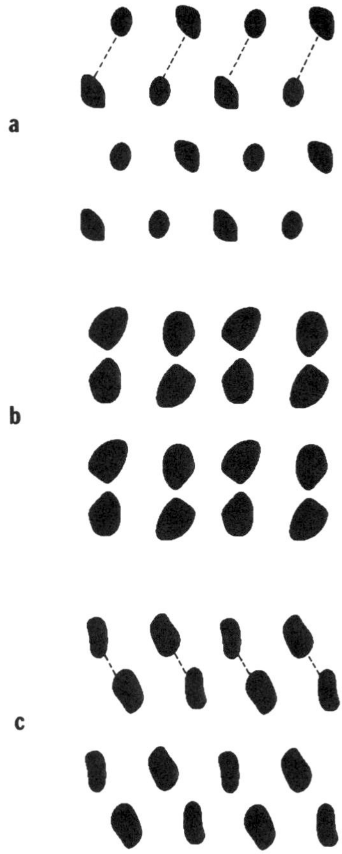

Figure 13. Sections through the three-dimensional map at three points in the model: (a) 3.0 nm above the central plane; (b) the central plane; and (c) 3.0 nm below the central plane. Note the change in apparent subunit position from one surface of the sheet (a) to the other (c) so that an approximate mirror image reversal of the pattern is produced.

VIII. RELATIONSHIP OF THE ELECTRON-MICROSCOPIC IMAGE TO X-RAY DATA FROM THE REACTION CENTER

The existence of well-ordered three-dimensional crystals of the *Rh. viridis* reaction center (Michel, 1982) have enabled a three-dimensional map to be produced at the atomic level of resolution (Diesenhofer et al., 1985) and have yielded interesting information concerning the organization of prosthetic groups in the central region of the reaction center (Diesenhofer et al., 1984). The details of the x-ray map can be compared to the map that has been derived by electron microscopy. Published descriptions of the reaction center suggest the existence of a rough cylinder with an elliptical cross-section in one region of the x-ray map (Diesenhofer et al., 1984), showing good agreement with the cross-section of the central map region revealed by electron microscopy (Figure 11). Deisenhofer et al. (1984) have reported, however, a maximum diameter for other regions of the reaction center of 7.0 nm, which we have not seen at any level of the EM map. They have also reported that the maximum dimension of the map is 14.0 nm. This value is much larger than the distance through which we have been able to trace the EM map.

There are, in general terms, two ways to explain the increased length of the reaction center in the x-ray map. Specimen dehydration, coupled with heavy metal staining, may have decreased the apparent length of the reaction center in the EM map. Because these factors do not affect the x-ray data, they may reflect the authentic dimensions of the reaction center. While the extraordinary (3-Å) detail of the x-ray map is beyond dispute, one noteworthy aspect of the *Rh. viridis* reaction center crystals is their very high (70%) solvent content. The inclusion of small ampiphiles may also have served to increase the "nonreaction center" content of these crystals. As noted earlier (Miller and Jacob, 1983), the 22.3 × 22.3 × 11.4 nm unit cell dimension of the three-dimensional crystals is much larger than 12.1 × 12.9 nm dimensions of the two-dimensional crystals, suggesting that the packing density of reaction centers in the three-dimensional crystals is much less than it is in the two-dimensional sheets. It is possible that the low protein density and the high solvent content of the three-dimensional crystals may have caused the unfolding of some molecular groups in the reaction center, resulting in a map that may be somewhat larger in some dimensions than the native membrane-bound reaction center.

IX. THE ACTUAL SYMMETRY OF THE REACTION CENTER CRYSTALLINE SHEET

In a recent study from our laboratory, deep-etching and rotary shadowing were used to examine the surfaces of two-dimensional crystalline sheets formed from the *Rh. viridis* reaction center. The results of this study suggested to us that the actual arrangement of four subunits within the unit cell corresponded to a symmetry of $P22_12_1$ (Miller and Jacob, 1985a). Although we assigned only P2 symmetry in this study, we believe that the higher symmetry is in fact correct for the arrangement of subunits within the sheet. Two major features are suggested by the shadowing work: (1) that there is a staggered arrangement of subunits at each surface of the sheet, and (2) that each of the four subunits should be shifted off the central plane of the sheet so that two were closer to one surface and two closer to the other.

Figure 13 illustrates that the first prediction is indeed fulfilled by the three-dimensional map. The contours of subunits near one surface are shifted in a staggered fashion, those near the membrane center are rectalinear, and those near the other surface are also shifted, although in mirror reverse, as required for the existence of a twofold symmetry axis within the plane of the membrane.

The second prediction of the shadowed work is not observed, however. This can best be explained by the EM technique used for the reconstruction. Our shadowing study used the deep-etching method in which frozen (hydrated) membrane samples are used for rotary replication. The three-dimensional reconstruction was carried out on negative-stained material in which membrane samples are dried down against a carbon film and dehydrated. During the drying process, the compression of the sheet against the film may have squeezed the sheet together so that all four subunits were compressed into roughly the same central plane. The drying process may explain, therefore, both the deviation from $P22_12_1$ symmetry and the reduced size of the reaction center when compared to the x-ray map.

X. HOW IS THE REACTION CENTER ORIENTED IN THE PHOTOSYNTHETIC MEMBRANE?

Although x-ray diffraction is clearly the method of choice to illustrate critical high-resolution details regarding the internal organization of the reaction center, one question which it cannot address at this time is the orientation of the reaction center in the *Rh. viridis* photosynthetic membrane. Although work is in progress at this time in several labs to produce higher-resolution EM maps of the *Rh. viridis* photosynthetic membrane, it might be worthwhile at this point to make a preliminary comparison between the EM map of the reaction center and my earlier data on the photosynthetic membrane.

When a graphic display of the *Rh. viridis* photosynthetic membrane is used to produce a three-dimensional map of one subunit, a direct comparison can be made between the two structures produced by Fourier methods from electron microscopy. Although the reaction center maps cannot be directly placed in the membrane subunit, the cylindrical shape of the reaction center, combined with its length (whether from x-ray or EM values) leaves us only one way to position the reaction center in the membrane—as the large central structure that protrudes from both surfaces of the photosynthetic membrane. The smaller masses which surround this central structure may then be interpreted as LH components clustered around the central reaction center, as many investigators have suggested (Welte and Kreuz, 1982; Stark et al., 1984; Miller, 1982; Jay et al., 1984).

If we place the reaction center in this structure, we can account for the protease sensitivity of the RC peptides (Jacob and Miller, 1983) and also allow the reaction center a position in which transmembrane electron flow can be accommodated. It does, however, raise an interesting question regarding the organization and orientation of individual subunits in the photosynthetic membrane. The hexagonal lattice of the native photosynthetic membrane stands in sharp contrast to the rectangular lattice of the reaction center sheets. There is complete agreement between x-ray and EM maps regarding the roughly elliptical cross-section of the individual reaction center. Is such an elliptical reaction center found within the center of each membrane subunit? And, if it is, are all such subunits oriented in precisely the same direction in the hexagonal array found in the photosynthetic membrane? Some have suggested, for example, that individual subunits within the photosynthetic membrane may take up a number of orientations that are then averaged together by Fourier techniques to produce the round shape that is normally reported for the central structure in the membrane (Engelhardt et al., 1983). In any case, the existence of a single asymmetric unit at the center of a structure with apparent sixfold symmetry presents a serious problem when one is trying to elucidate the structure of the membrane.

XI. MOLECULAR APPROACHES TO THE *RH. VIRIDIS* SYSTEM

Structural studies of the *Rh. viridis* system have been supplemented with molecular studies of the individual polypeptides. The amino acid sequence of the H subunit of the reaction center was determined by Michel et al. (1985), and more recently sequences from the M and L subunits have also been reported (Michel et al., 1986). The L subunit consists of 273 amino acids and has a molecular weight of 30,571, whereas the M subunit contains 323 amino acids with a molecular weight of 35,902. Similar values for the H subunit are 258 amino acids and a molecular weight of 28,345. (The fact that the H subunit is not the heaviest of the three stems from the fact that the names of the three subunits were assigned on the basis of migration behavior in sodium dodecyl sulfate(SDS)-polyacrylamide gels, where the H subunit displays an apparent molecular weight of 35 kd.) The combination of x-ray diffraction with polypeptide primary sequence has provided a detailed picture of the basic organization of the reaction center. The M and

L subunits, for example, each contain five membrane-spanning α-helical regions. The amino termini of the polypeptides are on the cytoplasmic side of the membrane, and together they display a degree of twofold symmetry (Diesenhofer et al., 1985). The cytochrome subunit is located entirely on the inner (periplasmic) side of the membrane. The H subunit is exposed on the cytoplasmic side, but contains an α-helical region that leads to its amino terminus on the periplasmic side of the membrane.

There is considerable primary sequence homology between the M and L subunits, and Michel et al. (1986) have suggested that these two polypeptides derive from duplication of a single ancestral gene. There also is a suggestion of homology between the L and M subunits and the 32-kd herbicide-binding protein associated with the higher plant photosystem II reaction center (Zurawski et al., 1982; Williams et al., 1983; Youvan et al., 1984).

The rapid extension of studies of the *Rh. viridis* system into the details of protein organization have provided us with the first detailed look at the organization of a photosynthetic reaction center. The *Rh. viridis* reaction center, it can be argued, is also now the best studied of all membrane proteins and will serve as a model against which techniques of protein structure prediction will be developed and calibrated.

XII. SOME UNSOLVED PROBLEMS

The crystals that we have studied have led to a three-dimensional view of a photosynthetic reaction center that can be placed into a specific position on a three-dimensional map of membrane subunits. In addition, x-ray crystallography has produced fascinating detail of the molecular groups within the reaction center (Diesenhofer et al., 1985). Nonetheless, we are led to ask further questions regarding the organization of the membrane itself. Our understanding of the *Rh. viridis* photosynthetic membrane has reached an unusual point. We now have an atomic level map of the RC and amino acid sequences of many of its components to add to our low-resolution three-dimensional map of the membrane itself. However, something is missing. Photosynthetic membranes in higher plants, algae, and bacteria contain a membrane-bound ATP synthetase, generally known as a coupling factor. Typically, coupling factors contain five polypeptides and have an aggregate molecular weight of about 400 kd (Merchant and Selman, 1985). They have been described for several species of photosynthetic bacteria, including *Rhodopseudomonas capsulata* (Baccharini-Melandri et al., 1970), *Rhodopseudomonas palustris* (Khananshvili and Gromet-Elhanan, 1983), *Rhodopseudomonas sphaeroides* (Baccharini-Melandri et al., 1970; Khananshvili and Gromet-Elhanan, 1983), and *Rhodospirillum rubrum* (Philosoph and Gromet-Elhanan, 1971). However, our current models of membrane organization for *Rh. viridis*, which contain only RC and LH components, do not allow room for coupling factor molecules. Indeed, the fact that isolated, purified photosynthetic membranes contain seven polypeptides, each now clearly identified, means that we have nothing "left over" to play the role of coupling factor. A recent attempt from our laboratory to reconcile the dimensions of the RC with the close apposition of membranes *in situ* concluded that there might be spatial problems in trying to fit a molecule as large as a coupling factor between two adjacent photosynthetic membranes (Miller & Jacob, 1985b).

Therefore, although we are developing an excellent structural model of the membrane, we are missing one of its most important functional components, and we lack an understanding of how this membrane is organized in the *functional sense.* All coupling factors are bound to integral membrane components that form an important part of the ATP-synthetase system (by convention, the soluble (400-kd) part of the complex is called CF_1 and the membrane-bound part is called CF_0) (Merchant and Selman, 1985). If coupling factors were associated with the hexagonal lattice of the *Rh. viridis* membrane, then we should observe CF_0 peptides associated with purified membranes. The fact that such peptides are not observed (Jacob and Miller, 1983, 1984; Jay et al., 1984) suggests that the coupling factor is missing from the membrane, at least where the hexagonal lattice is found.

The possibility cannot be discounted that the *Rh. viridis* coupling factor is localized in a nonlattice region contiguous with the regular lattices that we and workers in other laboratories have studied. If this were the case, then *Rh. viridis* would be similar to higher plants in which coupling factors have been shown to be restricted to the nonstacked regions of membrane and absent from stacked (grana) regions (Miller and Staehelin, 1976). However, we should then expect to find continuities between lattice and nonlattice membranes within the *Rh. viridis* cytoplasm. Such continuities have not been reported (Miller and Jacob, 1985b). The need for locating and identifying a coupling factor in this system cannot be underestimated. The possibility exists that *Rh. viridis* is substantially different from other photosynthetic organisms in the manner in which it synthesizes ATP. The emerging importance of this system as a model for photosynthesis makes it that much more important to determine if in fact the *Rh. viridis* system can serve as a general model for other species. Similarly, if *Rh. viridis* is substantially different from other photosynthetic bacteria, it will be exciting and instructive to detail and explore these differences.

XIII. THE INTEGRATION OF RESULTS

One of the advantages of this system, besides its relative simplicity and structural organization, is the opportunity that it presents to integrate information from different experimental techniques. We have already seen how protein chemistry, x-ray diffraction, and electron microscopy can be combined to describe an important biological structure. Future studies on this membrane will involve all of these techniques in addition to antibody labeling, protein cross-linking, and detergent solubilization. As complex as it is, the *Rh. viridis* system offers the possibility of integrating these several lines of work around a single biological problem. As work continues on these and other questions, it seems likely that the *Rh. viridis* photosynthetic membrane will prove an invaluable system in which to explore the molecular details of photosynthesis and to extend our understanding of membrane proteins.

ACKNOWLEDGMENTS

I am very much in debt to the many talents of Jules Jacob, who has assisted at every phase of this work. The Fourier imaging programs used for this study were kindly provided by Dr. Timothy Baker of Purdue University. This work was supported by a grant from the National Institutes of Health (GM 28799).

REFERENCES

Baccharini-Melandri, A., Gest, H., and San Pietro, A. (1970) J. Biol. Chem. *245:* 1224–1226.

Clayton, R. K., and Sistrom, W. R., eds. (1978) *The Photosynthetic Bacteria.* Plenum Press, New York.

DeRosier, D. J., and Klug, A. (1966) Nature (Lond.) *212:* 29–32.

Diesenhofer, J., Epp, O., Miki, K., Huber, R., and Michel, H. (1984) J. Mol. Biol. *180:* 385–398.

Diesenhofer, J., Epp, O., Miki, K., Huber, R., and Michel, H. (1985) Nature (Lond.) *318:* 618–624.

Engelhardt, H., Baumeister, W., and Saxton, O. (1983) Arch. Microbiol. *135:* 169–177.

Giesbrecht, P., and Drews, G. (1966) Arch. Mikrobiol. *54:* 297–330.

Jacob, J. S., and Miller, K. R. (1983) Arch. Biochem. Biophys. *223:* 282–290.

Jacob, J. S., and Miller, K. R. (1984) Biochem. Biophys. Res. Commun. *120:* 164–171.

Jay, F., Lambillotte, M., Stark, W., and Muhlethaler, K. (1984) EMBO J. *3:* 773–776.

Khananshvili, D., and Groment-Elhanan, Z. (1983) J. Biol. Chem. *258:* 3270–3275.

Merchant, S., and Selman, B. R. (1985) Photosynthesis Res. *6:* 3–31.

Michel, H. (1982) J. Mol. Biol. *158:* 567–572.

Michel, H., Weyer, K. A., Gruenberg, H., and Lottspeich, F. (1985) EMBO J. *4:* 1667–1672.

Michel, H. (1983) Trends Biochem. Sci. *8:* 56–59.

Michel, H., Weyer, K. A., Gruenberg, H., Dunger, I., Oesterhelt, D., and Lottspeich, F. (1986) EMBO J. *5:* 1149–1158.

Miller, K. R. (1979a) Sci. Am. *241*(4): 102–113.

Miller, K. R. (1979b) Proc. Natl. Acad. Sci. USA *76:* 6415–6419.

Miller, K. R. (1982) Nature (Lond.) *300:* 53–55.

Miller, K. R. (1985) Endeavor *9:* 175–182.

Miller, K. R., and Jacob, J. S. (1983) J. Cell Biol. *97:* 1266–1270.

Miller, K. R., and Jacob, J. S. (1985a) Eur. J. Cell Biol. *36:* 247–255.

Miller, K. R., and Jacob, J. S. (1985b) Arch. Microbiol. *142:* 333-339.
Miller, K. R., and Staehelin, L. A. (1976) J. Cell Biol. *68:* 30-49.
Parson, W. W. (1982) Annu. Rev. Biophys. Bioeng. *11:* 57-76.
Philosoph, S., and Gromet-Elhanan, Z. (1971) Eur. J. Biochem. *119:* 107-113.
Pucheu, N., Kerber, N. L., and Garcia, A. F. (1976) Arch. Microbiol. *109:* 301-305.
Stark, W., Kuhlbrandt, W., Wildhaber, I., Wehrli, E., and Muhlethaler, K. (1984) EMBO J. *3:* 777-783.
Thornber, J. P., Cogdell, R. J., Seftor, R. E. B., and Webster, G. D. (1980) Biochim. Biophys. Acta *593:* 60-75.
Welte, W., and Kreuz, W. (1982) Biochim. Biophys. Acta *692:* 479-486.
Williams, J. C., Steiner, L. A., Feher, G., and Simon, M. I. (1983) Proc. Natl. Acad. Sci. USA *81:* 7303-7307.
Youvan, D. C., Bylina, E. J., Alberti, M., Begusch, H., and Hearst, J. E. (1984) Cell *37:* 949-957.
Zurawski, G., Bohnert, H. J., Whitfield, P. R., and Bottomley, W. (1982) Proc. Natl. Acad. Sci. USA *79:* 7699-7703.

MOLECULAR CYTOCHEMISTRY OF FREEZE-FRACTURED CELLS:

FREEZE-ETCHING, FRACTURE-LABEL, FRACTURE-PERMEATION, AND LABEL-FRACTURE

Pedro Pinto da Silva

I. FREEZE-FRACTURE: MEMBRANE BILAYERS ARE SPLIT

The first platinum/carbon (Pt/C) replicas of freeze-fractured biological membranes showed smooth faces covered by particles. At the time, these were thought to represent proteins and nucleoproteins adsorbed to the membrane surfaces (Moor and Muhlethäler, 1963; Branton and Moor, 1964). Branton (1966) hypothesized that membranes were split during fracture and that the particles represented structures sequestered within the apolar matrix of the membrane. This proposal, which implied the existence of a lipid bilayer in membranes, was consistent with the later model of

Advances in Cell Biology, Vol. 1, pages 157–190.

ISBN: 0-89232-792-8

Danielli (1954) and with the unit membrane concept of Robertson (1961). It conflicted with the globular models of others, among them Sjöstrand (1963), Benson and Singer (1965), and Green and Perdue (1966). By covalently attaching ferritin molecules to both surfaces of erythrocyte membranes, I and Branton showed that the surface marker was not revealed by freeze-fracture; exposed ferritin molecules were seen only after the ice table surrounding the fracture face was lowered by sublimation ("etching"; see Figure 1) to reveal the true outer and inner surfaces of membranes (Pinto da Silva and Branton, 1970) (Figure 2). These experiments established the interpretation of freeze-fracture images and led to the synthesis of the globular and bilayer concepts of membrane structure: a bilayer continuum interrupted by protein intercalations (Pinto da Silva and Branton, 1970).

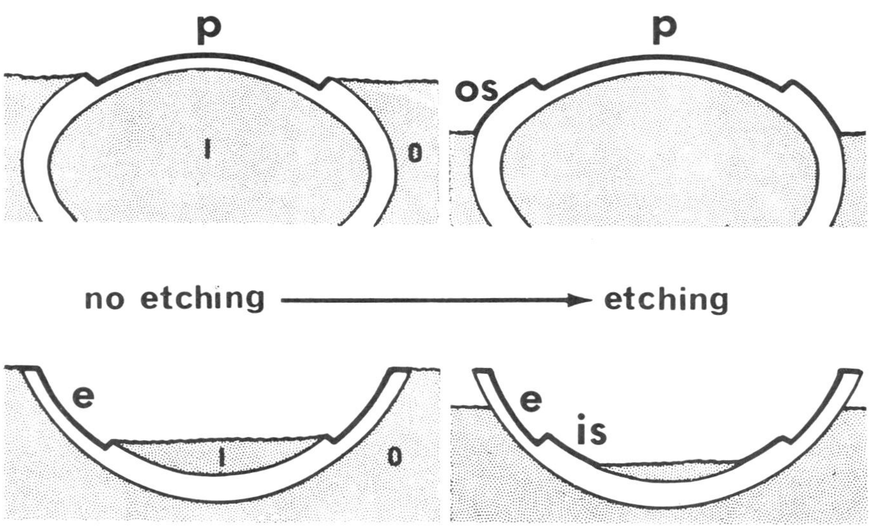

Figure 1. Fracture faces P and E are revealed as a result of membrane splitting by freeze-fracture. The outer (os) and inner (is) surfaces of the membrane are exposed by etching, i.e., lowering of the ice table by sublimation. If the membranes are frozen in a nonsublimable medium, the true surfaces cannot be seen (for the partition and labeling of transmembrane proteins, see Section III). Transmembrane proteins (not depicted) partition with protoplasmic or exoplasmic halves of the membrane. Therefore, surface receptors and antigens associated to transmembrane proteins may be labeled on protoplasmic halves (see Section III). (From Pinto da Silva and Branton, 1970.)

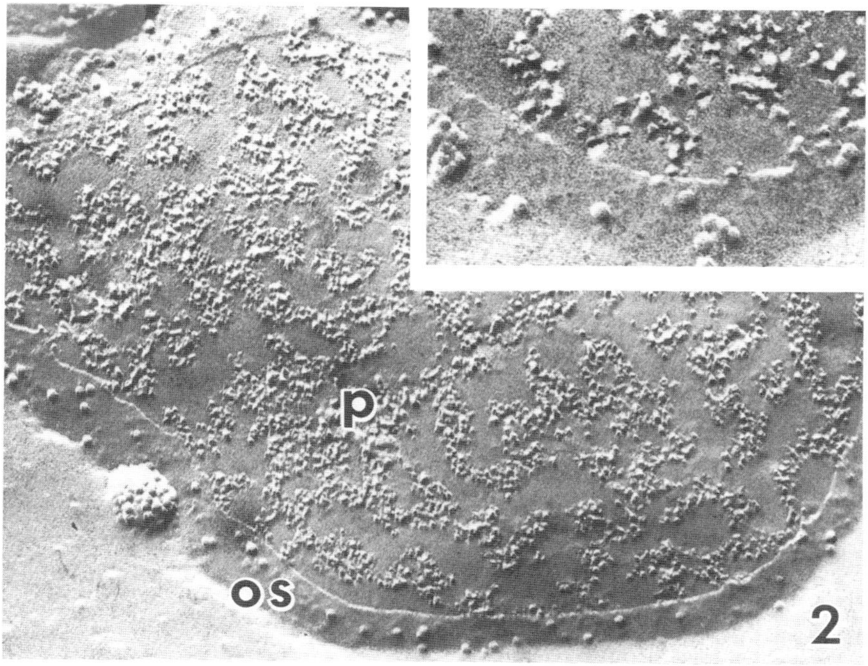

Figure 2. Freeze-etch cytochemistry shows that membranes are split by freeze-fracture. Ferritin was linked covalently to the outer and inner surfaces of erythrocyte ghost membranes. Because the membranes are split by freeze-fracture, ferritin molecules are not seen on the fracture face (p; see also Figure 1) but are visible over the surfaces exposed by etching (os). ×70,000; inset, ×200,000. (From Pinto da Silva and Branton, 1970.)

II. FREEZE-ETCHING: FREEZE-FRACTURE CYTOCHEMISTRY STARTS

The ferritin experiments showed also that etching could be used to relate the distribution of the intramembrane particles on the fracture face to that of a recognizable label of specific molecules expressed at the etched membrane surfaces.

In the first experiments of freeze-etch immunochemistry (Pinto da Silva et al., 1970, 1971) we used anti-human IgG (immunoglobulin G) conjugated to ferritin to label human erythrocytes incubated with anti-A antibodies. We showed that in human erythocytes the distribution of the A antigen coincided with that of the intramembrane particles. This finding implied that the particles must represent intercalated structures that reach the outer surface of the membrane. Because the particles partition preferentially with the inner (protoplasmic) half of the membrane, we proposed that they represented transmembrane proteins (Pinto da Silva et al., 1970, 1971). Subsequent freeze-etch labeling experiments extended and confirmed these observations by ascribing to the membrane particles a variety of lectin receptors (Figure 3), viral antigens, and negative charges (these labeled the intramembrane particles at both outer and inner surfaces of the membrane) (Tillack et al., 1972; Marchesi et al., 1972; Pinto da Silva et al., 1973; Pinto da Silva, 1973; Pinto da Silva and Nicolson, 1974). Etching cytochemistry

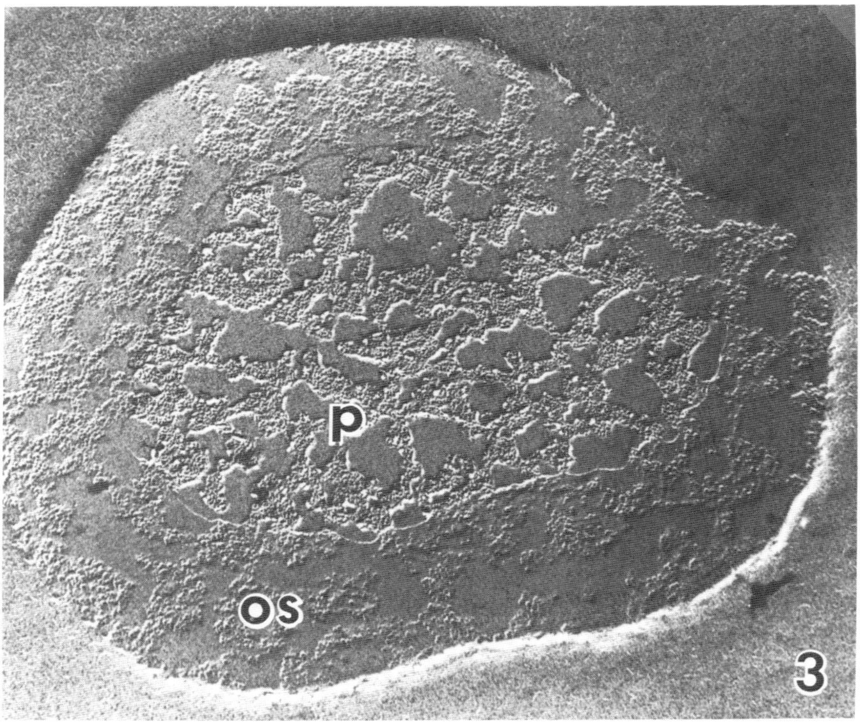

Figure 3. Human erythrocyte ghost membrane labeled with ferritin–concanavalin A. The pattern of aggregation of the label on the outer surface (os) closely resembles the random network of aggregated membrane-intercalated particles on the protoplasmic fracture face (p). The pattern of distribution of ferritin–concanavalin A is both contiguous and continuous with that of the membrane-intercalated particles. ×40,000. (From Pinto da Silva and Nicolson, 1974.)

required freezing in the absence of cryoprotectants to expose the true surface of membranes. In addition, it required the aggregation of the membrane particles into recognizable domains larger than the labeling molecules. The presence of cytoplasm and of organelle matrices prevented the exposure of true membrane surfaces even when the high speed of freezing rendered unnecessary the impregnation with a (nonsublimable) cryoprotectant.

In the following sections I describe three alternative techniques for the cytochemical characterization of the components (membrane associated or not) that are exposed by freeze-fracture; the first, *fracture-label* permits the labeling of plasma and intracellular membranes (as well as cytoplasm and nucleoplasm) after freeze-fracture; the second, *fracture-permeation* explores the compactness of cytoplasmic and nucleoplasmic matrices in glutaraldehyde fixed cells; the third, *label-fracture* relates directly the surface labeling of a membrane to its freeze-fracture morphology.

III. FRACTURE-LABEL: *IN SITU* CYTOCHEMISTRY OF MEMBRANE HALVES

A. Rationale and Goals

Freeze-fracture splits and exposes plasma membranes and intracellular membranes; in addition, cross-fracture exposes components of both cytoplasm and nucleoplasm. As conventional freeze-fracture is performed under vacuum, the molecules it exposes cannot be labeled because the specimens are frozen and inaccessible inside a vacuum chamber. We reasoned that if glutaraldehyde-fixed cells were freeze-fractured while immersed in liquid nitrogen and then thawed, the ultrastructure might still be recognizable and cellular components be available for labeling. The distribution of the label would then be observed either in thin sections or in Pt/C replicas of these freeze-fractured, cytochemically labeled specimens. The experiments that followed met our expectations and led to the development of fracture-label (Pinto da Silva et al., 1981a,b,c).

Fracture-label allows the cytochemical characterization of biological specimens *after fracture-fracture.* To date we have shown that with fracture-label it is possible to (1) identify transmembrane proteins and establish their pattern of distribution on the protoplasmic half of split membrane; (2) determine the partition and topology of lipids and proteins associated with the protoplasmic and exoplasmic "halves" of plasma and intracellular membranes; (3) study the intracellular traffic of specific membrane components and investigate the topochemistry and compartmentalization of intracellular membranes; and (4) label receptor sites exposed by cross fracture of the nucleoplasm and the cytoplasm.

B. Technique

For fracture label, glutaraldehyde-fixed, glycerol-impregnated specimens are freeze fractured in liquid nitrogen, thawed, deglycerinated, and then labeled. The technical details of fracture label are reviewed elsewhere (Pinto da Silva et al., 1986; Pinto da Silva, 1986a,b), as well as in the original references.

There are two variants of fracture-label: (1) in *thin-section fracture-label* (Pinto da Silva et al., 1981b,c) the specimens are crushed under liquid nitrogen, into fine fragments, with a tissue homogenizer and a narrower (mismatched) glass pestle (Figure 4). After thawing and labeling, the suspension of fragments is processed for thin-section electron microscopy; (2) in *critical-point–drying fracture-label* (Pinto da Silva et al., 1981a) the frozen specimens are also immersed in liquid nitrogen, then are fractured with a cooled scalpel, and the specimen fragments are thawed, labeled, and critical-point-dried. These fracture-labeled fragments are then shadowed with Pt/C, and the replicas observed with a transmission electron microscope.

C. Preservation of Ultrastructure and Reorganization of Membrane Components after Fracture

On observation of fracture-labeled specimens it became immediately apparent that thawing of glutaraldehyde-fixed, glycerol-impregnated specimens does not affect their ultrastructure (figure 5). I believe that one of the main reasons why post-fracture labeling techniques were not attempted over the first 20 years of freeze-fracture was the assumption that thawing would result in the obliteration of cell ultrastructure. This assumption was reinforced by the microtrabecular theories of cytoplasmic structure that viewed the cytoplasm as a relatively open and delicate network containing all cytoplasmic proteins and nucleoproteins, bathed in an aqueous phase of ions and small metabolites (Guatelli et al., 1982; Porter et al., 1982). Such delicate structures would not resist the brutal disruption caused by the formation, on thawing, of large crystals even in a glycerol-enriched aqueous phase. Instead, observation of fracture-labeled specimens show that routine impregnation of specimens in glycerol, followed by freezing, thawing, and deglycerination, does not imply a decreased preservation of the ultrastructure.

Ultrastructural observation of the fractured "halves" of the membranes frequently reveals a puzzling profile: instead of a single dense line expected from split membranes[1] (i.e., one-half of the trilaminar "unit membrane"), we often see an interrupted trilaminar profile (Pinto da Silva et al., 1981b,c; Barbosa and Pinto da Silva, 1983) (Figure 6). Does this observation question the location of the plane of fracture? I do not think so. If freeze-fractured

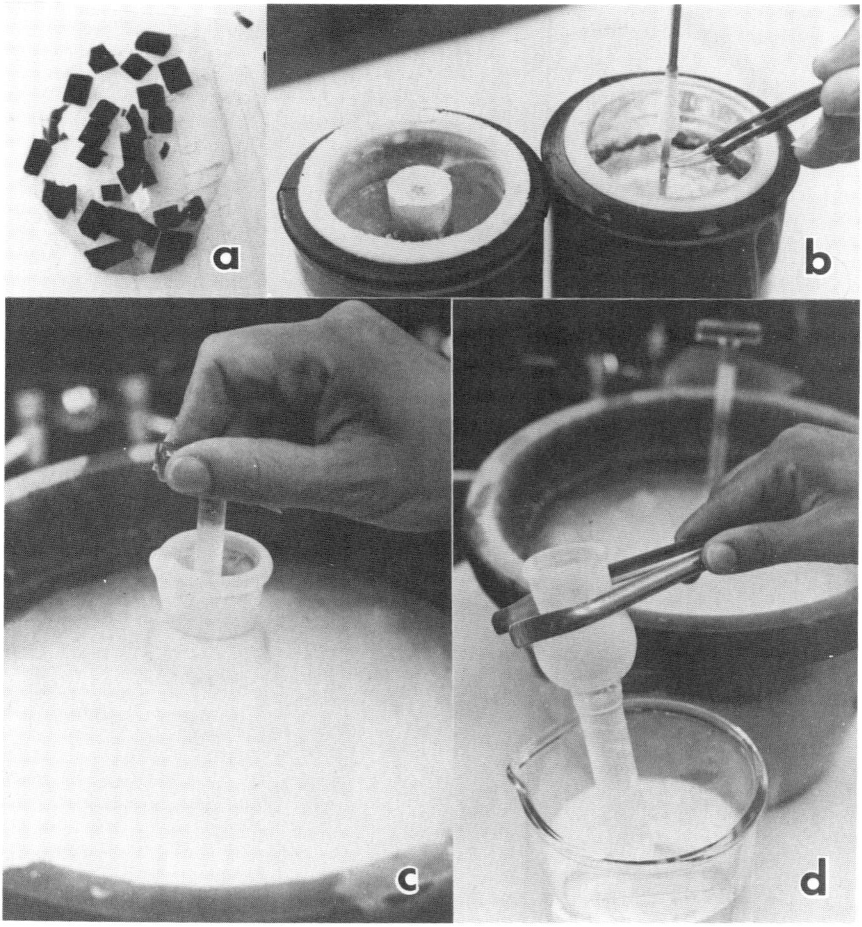

Figure 4. Technical aspects of thin-section fracture-label. (a) Cut "gels" of cells-BSA. (b) Freezing of "gels" in Freon 22 (*left*) and in liquid nitrogen. (c) Freeze fracture of gels in liquid nitrogen. (d) Thawing of freeze-fractured specimens. (From Pinto da Silva et al., 1986.)

specimens are not thawed in an aqueous environment but instead are freeze-substituted in osmium/acetone, fractured membranes appear as a single leaflet, frequently difficult to distinguish from the underlying cytoplasm (Pinto da Silva et al., 1981b,c). We have therefore proposed that, during thawing in an aqueous solution, membrane components (probably lipids) reorganize into bilayered structures. Because membrane "halves"

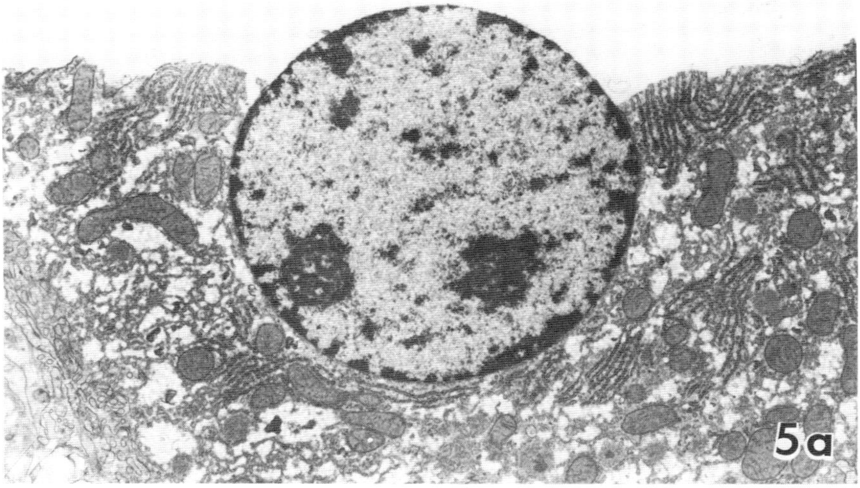

Figure 5. Cell ultrastructure after freeze-fracture and thawing. (a) Rat hepatocyte, thin-section view. ×6,500. (From Pinto da Silva et al., 1981b.) (b) Pt/C replica of a rat pancreatic acinar cell, after critical-point drying. ×40,000. (From Pinto da Silva et al., 1981b.) *Abbreviations*: PM_p, protoplasmic half of the plasma membrane; N, nucleus; SG, secretory granule; Er, endoplasmic reticulum. ×40,000. (From M. Torrisi and P. Pinto da Silva, unpublished observations.)

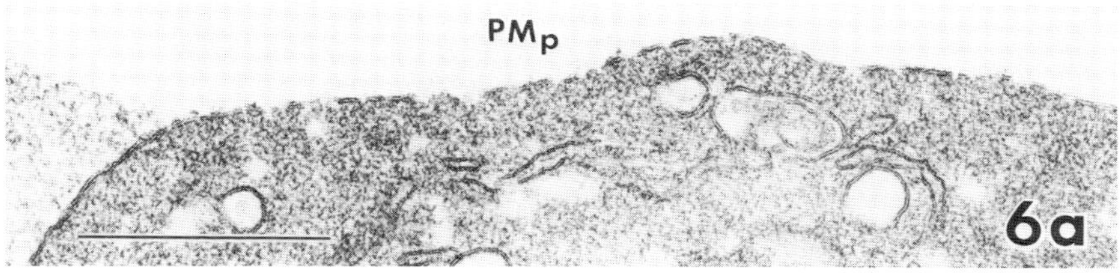

Figure 6. Ultrastructure of freeze-fractured membrane halves after thawing in glutaraldehyde solutions. (a) Human leukocyte (PM_p, protoplasmic half of the plasma membrane). ×48,000. (From Pinto da Silva et al., 1981b.) (b) Human erythrocyte membrane, protoplasmic half. ×67,000. (From Pinto da Silva and Torrisi, 1982.) See text for explanation and discussion.

cannot reconstitute a complete membrane with the original surface, a meshwork bilayered structure results that, in thin sections, may appear as an interrupted "unit membrane."

D. Identification of Transmembrane Proteins

I summarize here studies that use labeling of the protoplasmic membrane halves to detect transmembrane proteins. These studies addressed first the partition during fracture of the principal transmembrane proteins [band 3 (the anion transport protein) and glycophorin] in human erythrocyte membranes (Pinto da Silva et al., 1981b; Pinto da Silva and Torrisi, 1982). Fracture-labeling of erythrocyte membranes with concanavalin A (the Con A receptor is located on band 3) revealed that most of its surface receptors partition with the protoplasmic face. Fracture-labeling with wheat germ agglutinin (WGA) showed also that, while most WGA receptors (sialic acid; *N*-acetylglucosamine) partitioned with the exoplasmic half of the membrane

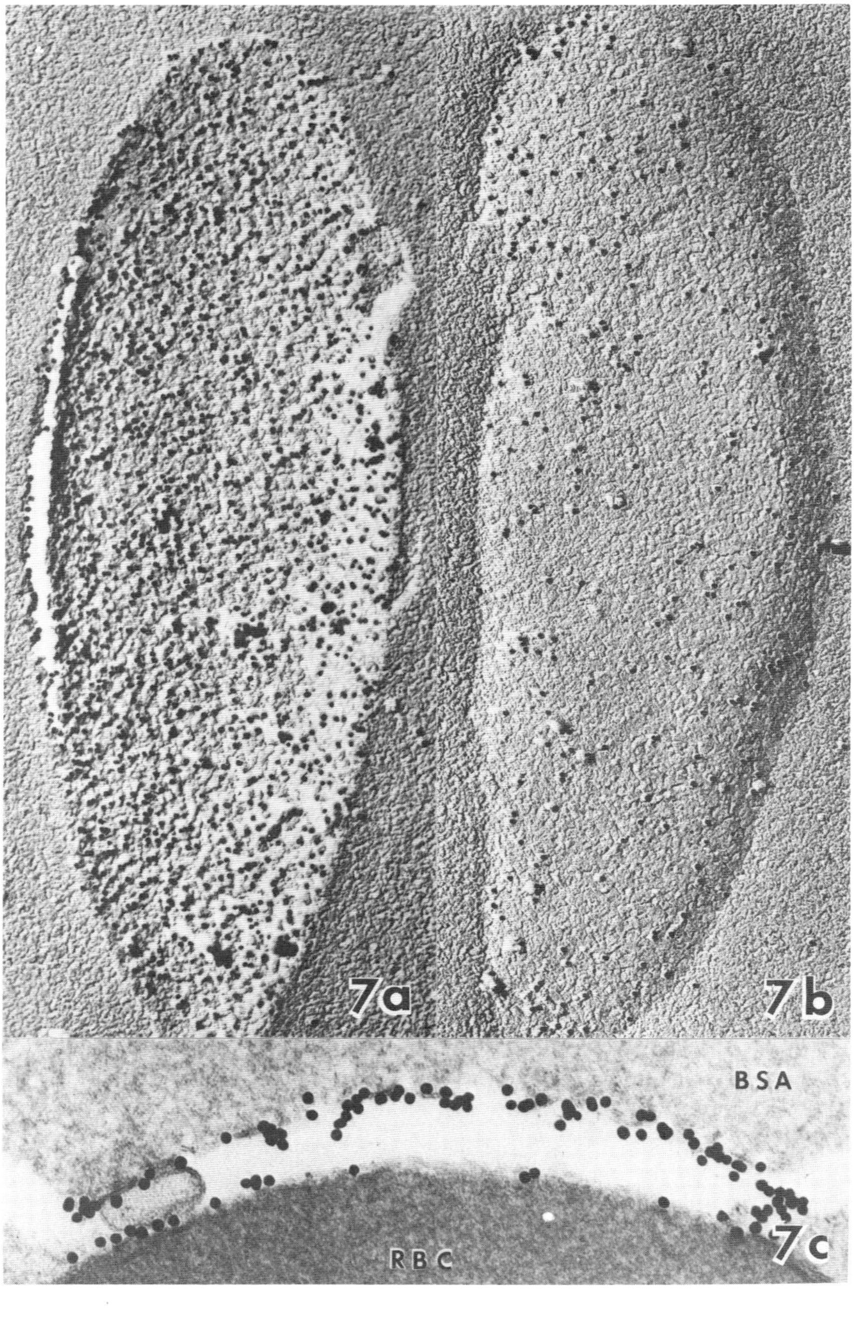
7a
7b
BSA
RBC
7c

(Figure 7a), a significant minority was always associated with the protoplasmic half (Figure 7b,c). These results, observed with the two variants of fracture-label demonstrated that, on fracture, transmembrane glycoproteins could be pulled across the outer half of the membrane along with their heterosaccharide chains. Lectin receptors on these chains could then be labeled on the protoplasmic half of the membrane. In addition to revealing the process of fracture of nonbilayered regions of the membrane, this result provided a method for the identification of the transmembrane proteins that, on fracture, partition with the protoplasmic "half" of the membrane. Critical-point-dried replicas of protoplasmic membrane halves revealed the patterns of distribution of these proteins along the plane of the membrane (Figure 7a,b).

When human T lymphocytes are fracture-labeled with WGA (Torrisi and Pinto da Silva, 1982) exoplasmic halves are densely marked, but labeling of protoplasmic halves displays striking variations—from virtually no label to very dense labeling (Figure 8). These variations were not the result of differences in accessibility of the cells to the marker; in critical-point-dried preparations many cells can be observed in a single fracture face of the gels (fixed cells are suspended in 25–30% solutions of bovine serum albumin (BSA), which is then cross-linked into a stiff "gel" by glutaraldehyde). In these preparations the differences in labeling density could be observed in protoplasmic faces lying side by side. We could, therefore, conclude that (1) WGA labeling of the protoplasmic halves reflects the existence of WGA-binding, transmembrane sialoproteins, and (2) human T lymphocytes are a heterogeneous population with respect to the presence of transmembrane sialoproteins. This work showed, therefore, that fracture-labeling can be used to search and characterize cell subpopulations based on differences in the expression of transmembrane proteins.

In addition to the search for individual variations of transmembrane proteins, fracture label can be used to detect their regionalization along the plane of the membrane. We used boar sperm as an outstanding example of a highly polarized cell (see Friend, 1982; Primakoff and Myles, 1983).

Figure 7. Labeling of freeze-fractured erythrocytes by WGA–ovomucoid–colloidal gold. (a,b) Pt/C replicas of critical-point-dried cells. Both, ×42,000. (c) Thin-section fracture-label. ×100,000. WGA receptors (glycophorin) partition preferentially with the exoplasmic half (panel a; panel c, *top*) of the membrane; the protoplasmic membrane half is sparsely labeled (panel b; panel c, *bottom*). *Abbreviations*: BSA, bovine serum albumin; RBC, red blood cells. (From Pinto da Silva and Torrisi, 1982.)

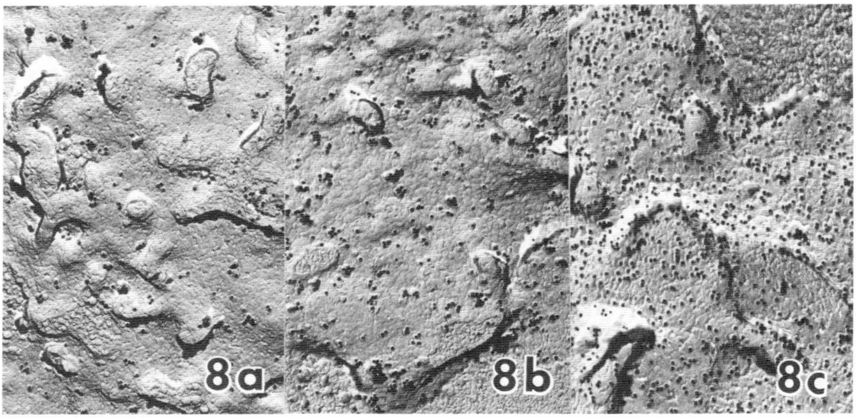

Figure 8. Pt/C replicas of fractured critical-point-dried purified T cells. The WGA label on the P faces is uniformly distributed and ranges from virtual absence to sparse (a), moderate (b), and high density (c). All, ×20,000. (From Torrisi and Pinto da Silva, 1982.)

Again, WGA and Con A fracture-label showed that many receptors partition with the protoplasmic face over the sperm head as well as over the midpiece and principal piece of the sperm tail (Aguas and Pinto da Silva, 1983, 1984). Critical-point-dried preparations revealed that, over the head, the density of WGA receptors is higher (about twofold) in the region of the plasma membrane that overlays the acrosome. This preferential distribution coincides with that of the larger intramembrane particles observed in conventionally freeze-fractured membranes. The finding of transmembrane glycoproteins in the sperm tail membrane is consistent with models that postulate transmembrane proteins of flagella as bearing motility-coupled cell surface receptors (Bloodgood, 1981).

Recently, Torrisi and Buonatti (1985) used fracture-label to detect Sindbis virus membrane proteins in infected chick embryo fibroblasts. In intracellular membranes the viral components partitioned always with the exoplasmic half of the membrane. In contrast, Sindbis virus glycoproteins partitioned preferentially with the protoplasmic halves of the plasma membranes and were frequently distributed in clusters related to the underlying nucleocapsids. This partition appears related to the binding to the nucleocapsid that takes place during budding.

The experiments summarized above give an idea of the unique capacity of fracture-label to detect transmembrane proteins when the receptors that they normally expose at the outer surface can partition, on fracture, with the protoplasmic half of the membrane. Unfortunately, this often is not the case, at least with Con A and WGA receptors (Pinto da Silva et al., 1981d). Yet, in most instances, intramembrane particles partition preferentially with protoplasmic halves. Is it that lectin receptors of cell surfaces are not, in most cases, represented by intramembrane particles? Recent work in Fisher's laboratory shows that, at least in erythrocyte membranes, the covalent backbone of integral proteins is not broken during freeze-fracture (see Fisher and Yanagimoto, 1986; also Fisher's paper in this volume). I believe that the answers to the foregoing question will have to wait for the labeling of other lectin receptors and of other cell surface antigens, coupled directly or indirectly (protein A) to colloidal gold, to explore in detail the partition of many important antigens, receptors, and epitopes on the cell surface. Still, many avenues lie open to research with fracture-label: For instance, are B cells also heterogeneous populations with respect to transmembrane sialoproteins? And, if so, are there differences between normal and leukemic B cells? Could partition have diagnostic and/or prognostic value? Clearly, fracture-label, as a new technique, affords new parameters that can now be used to reassess problems in the biology and, in particular, in the pathology of cells.

E. Glycolipids are Restricted to the Exoplasmic Half of the Plasma Membrane

Fracture-label experiments also established, for the first time, the topology of membrane glycolipids. This is a problem that eluded biochemists for some time but for which no solution was found; there was some expectation that, similarly to glycoproteins, the sugar moieties in glycolipids should be exposed at the exoplasmic surfaces.

We used *Acanthamoeba castellani*, a cell whose plasma membrane is devoid of glycoproteins with receptors for Con A (Barbosa and Pinto da Silva, 1983). Freeze-fracture of its plasma membrane shows faces virtually barren of particles and rugosities (Bowers, 1980), the only plasma membrane that I have seen with this appearance. The membrane of *A. castellani* contains a glycolipid (Korn et al., 1974; Korn and Wright, 1973) that binds Con A. Fracture-label of Con A receptors showed that virtually all of the label was associated with the exoplasmic half of plasma membranes or of intracellular membranes. To confirm the exoplasmic topology of glycolipids, we labeled the cytoplasmic surface of the plasma membrane, rendered accessible to the label by freeze/thaw cycles. We found that no label was associated with the cytoplasmic surface of either the plasma membrane or that of intracellular organelles (these displayed heavy label on their internal, i.e., topologically external, exoplasmic surfaces).[2]

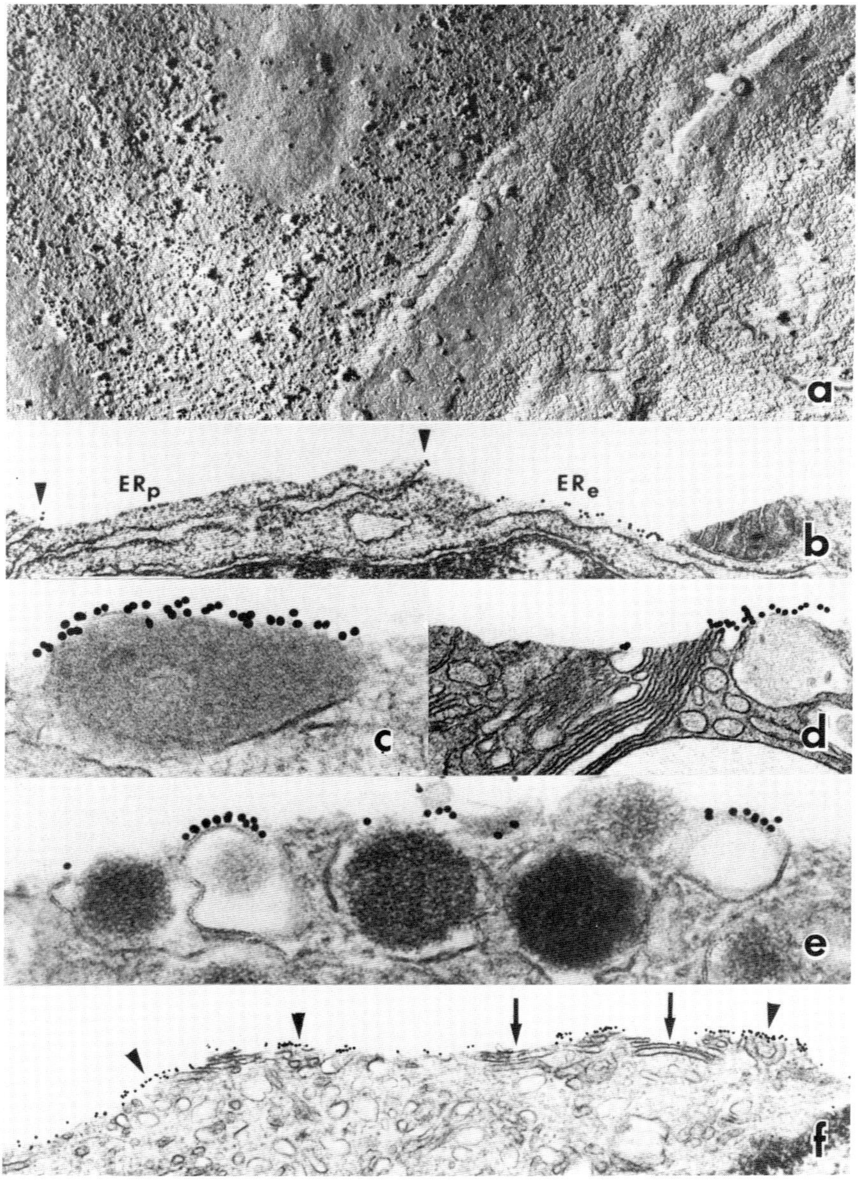

Figure 9. Thin-section fracture-label of lectin receptors of intracellular membranes. (a) WGA labels the exoplasmic half of the plasma membrane but fails to label the endoplasmic reticulum membrane. ×26,000. (b) Con A labels strongly the exoplasmic halves of endoplasmic reticulum

F. Intracellular Distribution and Traffic of Membrane Components

The course of fracture follows the path of least resistance. Frequently, this path cross-fractures cells, exposing not only elements of the cytoplasm and nucleoplasm but also faces of intracellular membranes. The components—membranous or not—thus exposed can be labeled. By thin-section of fracture-label we can observe the cellular ultrastructure underneath the plane of fracture and know which organelle or cytoplasmic region is labeled. With critical-point-drying fracture-label we are given, in turn, a two-dimensional view of the labeling along the entire plane of fracture. The two types of observation are complementary. We have now used fracture-label to illustrate these newly found possibilities of cytochemical access to intracellular membranes.

We fracture-labeled secretory and nonsecretory rat tissues and human leukocytes with Con A and WGA (Pinto da Silva et al., 1981b,d; Torrisi and Pinto da Silva, 1984). Both lectins labeled the plasma membrane, with most of the label associated with exoplasmic faces (Figure 9a). However, while Con A also labeled fractured endoplasmic reticulum and nuclear envelope membranes, WGA did not (Figure 9a,b). WGA labeled weakly the Golgi apparatus, with the exception of its transmost elements, which—as the membranes of the lysosomal system and of secretory vesicles—were strongly labeled (Figure 9c–e). These results provided direct evidence of the intracellular traffic of membrane glycoproteins. Con A labeled strongly the peroxisomes and the endoplasmic reticulum, membranes rich in mannose residues. Many of these residues are "clipped" before the glycoproteins reach the Golgi apparatus. The addition of terminal sugars (sialic acid; *N*-acetyl-D-glucosamine) occurs at the transmost cisternae of the Golgi as is illustrated by WGA labeling of its *trans* elements of the Golgi and lysosomal membranes (Figure 9c,d). In other instances labeling of the Golgi apparatus can be heavy (Figure 9f). The lack of labeling of the endoplasmic reticulum and nuclear envelope membranes by WGA showed that there is no detectable "backflow" of fully glycosylated products from the Golgi

---

Figure 9—continued
membranes (ER_e) [as well as cross-fractures (*arrowheads*)]. ×35,000. (c–e) WGA fracture label of a lysosomal membrane, Golgi apparatus, and insulin granule membranes, respectively. a, ×78,000; d, ×49,000; e, ×63,000. (f) fracture label of anti-Sindbis envelope glycoproteins on fracture Golgi membranes in BHK cells, 3 hours after infection. ×26,000. (Panels a, b, and e from Pinto da Silva et al., 1981d; panels c and d from Torrisi and Pinto da Silva, 1984; panel f from Torrisi and Bonatti, 1985.)

apparatus into the endoplasmic reticulum or nuclear envelope membranes. It would therefore appear that the previous identification of fully glycosylated products in these membranes was the result of cross-contamination during membrane fractionation. In fracture-label, cross-contamination does not appear to be significant: only components from the fractured membranes can be sources of contamination, and these are a tiny fraction relative to the bulk of unfractured membranes; moreover, even if membrane components were liberated from the fracture face, the areas of fractured membrane susceptible to contamination are tiny as compared to the total areas to which the contaminant can adhere nonspecifically (e.g. glass containers, BSA gels). The observation of complementary (i.e., protoplasmic and exoplasmic) aspects of labeling in "cracked" cells shows that cross-contamination does not appear to occur: frequently we see most or all of the label associated with one of the fracture faces; should there be a significant cross-contamination, it would tend to equalize the labeling on both fracture faces. Note, however, that contamination can exist; I have referred earlier to the adherence of freed glycogen granules to the protoplasmic faces in *A. castelannii* and to the liberation of membrane lipids (Section III.E).

In addition to detecting intracellular receptors, fracture-label in particular critical-point-drying fracture-label—can detect variations in their distribution along the membrane. Fracture-label of the acrosomal membrane in boar spermatozoa showed that this is a membrane remarkably poor in glycoconjugates (Aguas and Pinto da Silva, 1985). The acrosomal membrane displayed a regional accumulation of Con A receptors over the equatorial segment of the exoplasmic membrane half.

G. Topology of Glycoproteins in the Nucleoplasm

The two main compartments of the nucleoplasm observed in conventional thin sections—euchromatin and heterochromatin—are easily distinguished in cross-fractured nuclei after critical-point and Pt/C replication. The heterochromatin is more compact and appears smoother than the euchromatin, which is less compact and displays a rougher texture. We have used Con A and *Ulex europaeus* lectin to show that receptors to these lectins exist within the nucleoplasm, where they are mostly associated with the euchromatin regions (Kan and Pinto da Silva, 1986) (Figure 10). As trypsin-treated cells are not labeled, the receptors must be associated to glycoproteins. This is a fairly new area of research where little is known. Some intranuclear glycoproteins—the high-mobility-group (HMG) proteins—are glycosylated and appear to be associated with active chromatin (Reeves et al., 1981; Weisbrod et al., 1980; Levy-Wilson et al., 1979; Bakayev et al., 1979). Other cytochemical techniques could, in principle, have been used to arrive at our findings. An attempt with frozen sections did not, however,

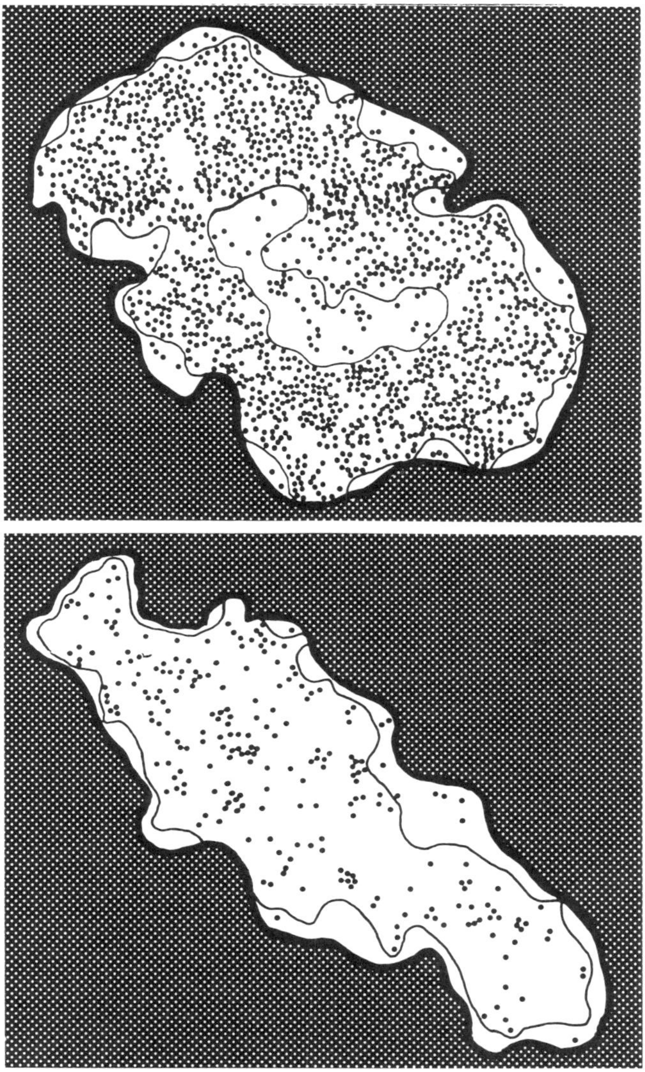

Figure 10. Tracings of cross-fractured nuclei from replicas of duodenal collumnar cells after critical-point-drying fracture label. Labeling by *Ulex europaeus* agglutinin-I. *Top*: experimental. *Bottom*: L-fucose control. (From Kan and Pinto da Silva, 1986.)

have enough resolution (Seve et al., 1984); labeling of polymer-embedded sections has not yet been tried. I should note here that one possibly crucial advantage of fracture-label is that the receptors to be detected are exposed to the label to a higher degree than those in plastic-embedded sections, where only those at the actual interface may be sterically accessible.[3]

IV. FRACTURE-PERMEATION: ASSESSMENT OF THE COMPACTNESS OF THE CYTOPLASM

For fracture-permeation (Barbosa and Pinto da Silva, 1986a,b) glutaraldehyde-fixed cells are freeze-fractured and, after thawing, are immersed into very concentrated solutions of a macromolecular tracer, e.g., ferritin. According to the degree of packing of the molecules in the cytoplasm after glutaraldehyde fixation, the tracer will permeate (or not) the fractured cytoplasm. In initial model experiments done with glutaraldehyde-fixed BSA matrices, we showed that ferritin was able to permeate gels of 10% and 15% BSA but could not permeate gels of 20% and 25% BSA (Figure 11). When used to probe the cytoplasmic compactness of cells, fracture-permeation showed that the cytoplasm of mature human neutrophyls (after glutaraldehyde fixation) is not permeated by ferritin (Barbosa and Pinto da Silva, 1985) (Figure 12a). This is of particular interest because these cells were used by Porter and his co-workers (see Pryzwansky et al., 1983) to show the microtrabecular lattice as an open scaffold whose dimensions should permit free permeation and flow of ferritin.

With fracture-permeation we showed that the compactness of the cytoplasm, as judged by the permeation by native ferritin of glutaraldehyde-fixed, cross-fractured cells, varies with cellular activity, during development, and in different physiological conditions (Barbosa and Pinto da Silva, 1986a). Resting human lymphocytes are not (or very scantly) permeated by ferritin. In contrast, after activation by phytohemaglutinin, their cross-fractured cytoplasm is massively permeated by ferritin. We have also found that the compactness of the cytoplasm in *Phytophthora palmivora* (a fungus) varies during its life cycle; while the glutaraldehyde-fixed cytoplasm of sporangia is freely permeated by ferritin, that of zoospores and of cysts is compact and is not permeated. Upon increase of cellular activity, i.e., during germination of cysts, the cytoplasm of the cysts is permeated, with the exception of the germinating tubes, which display a compact cytoplasm.

In cardiac and skeletal muscle, permeation by ferritin creates patterns that can be used to characterize each stage of contraction (Barbosa and Pinto da Silva, 1986b): while ferritin cannot permeate into shortened sarcomeres from muscle incubated in "rigor" solutions, it permeates the I band of sarcomeres at relaxed length (Figure 12b) and the I band and H

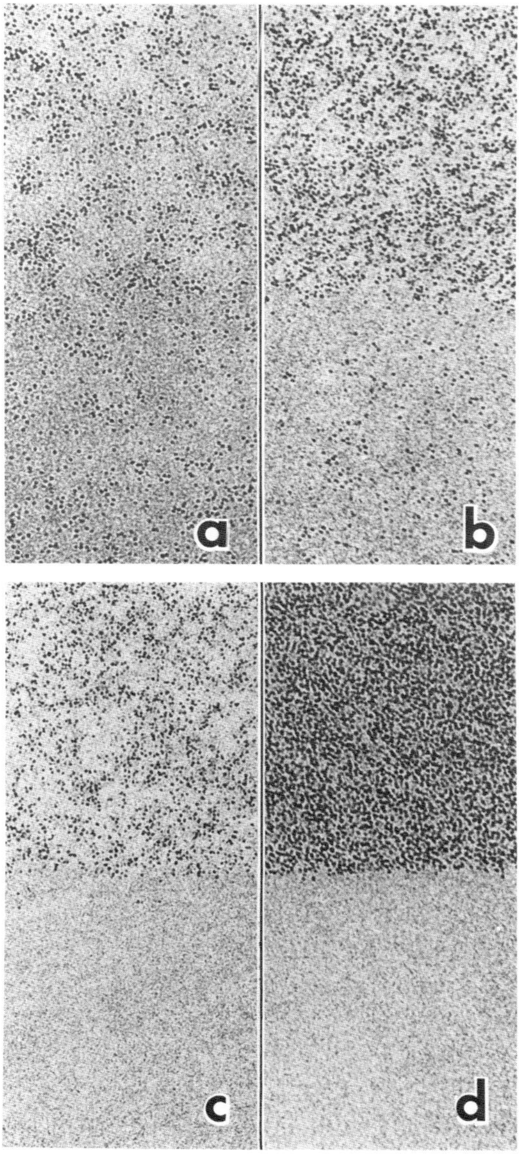

Figure 11. Permeation of ferritin molecules into freeze-fractured BSA "matrices." The matrices were made of bovine serum albumin (BSA) cross-linked by glutaraldehyde. Ferritin permeated only matrices made from (a) 10% or (b) 15% (weight/volume) BSA solutions. (c) Matrix made from 20% BSA solution. (d) Matrix made from 25% BSA solution. Fracture faces on matrices from 10% BSA solution were irregular and hard to distinguish in unstained thin sections. All, ×50,000. (From Barbosa and Pinto da Silva, 1986.)

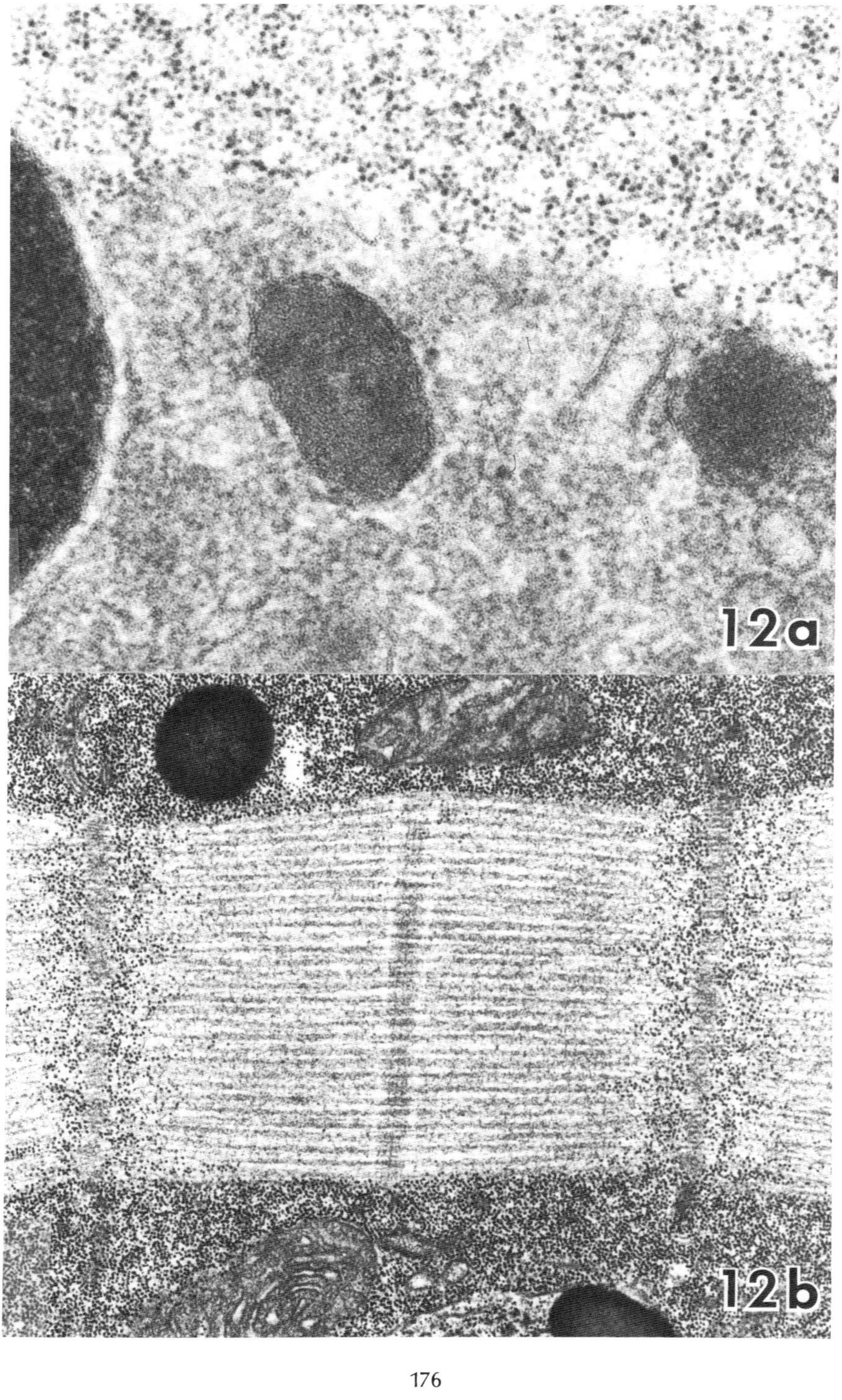
12a
12b

zone of the longest (presumably stretched) sarcomeres. These results provide a control for fracture-permeation because the patterns of permeation are those expected from the intramolecular (myosin, actin) distances observed in ultrastructural and x-ray-diffraction studies. In sarcomeres with a length intermediate between that of sarcomeres from muscle incubated in rigor solutions and those in the relaxed state, i.e., probably in the contracted state, ferritin permeates into the A band, thereby indicating irregularities in the interactions between actin and myosin during contraction. These irregularities could derive from slight asynchronies during filament sliding. We believe that fracture-permeation will have applications not only in cell biology but also in cell pathology because it provides a new parameter for the assessment of compactness of cytomatrices after glutaraldehyde fixation.

V. LABEL-FRACTURE: COINCIDENT IMAGES OF SURFACE LABELING AND EXOPLASMIC FACES

A. Rationale and Development

Ever since freeze-fracture revealed membrane faces full of an overwhelming and enigmatic variety of structural detail, their direct cytochemical characterization became a tantalizing objective. We have seen (above) how freeze-etching cytochemistry provided the first method of relating intramembrane particles to the distribution of surface receptors and antigens. Freeze-etching led to the cytochemical characterization of the topology and chemistry of the entities represented by the intramembrane particles. I described how, some 10 years later, fracture-label allowed cytochemical access to all plasma and intracellular membranes *after fracture*, leading to the identification of transmembrane proteins and their mode of partition, as well as to labeling of nonmembrane components. Yet, none of these techniques was able to provide images combining high-resolution surface labeling with conventional freeze-fracture. This has now been achieved by label-fracture (Pinto da Silva and Kan, 1984).

Figure 12. Fracture permeation. (a) Ferritin does not penetrate the cytoplasm of a glutaraldehyde-fixed, cross-fractured neutrophil. ×135,000. (From M. L. F. Barbosa and P. Pinto da Silva, unpublished observations.) (b) Fracture-permeation of cardiac muscle. Sarcomere at relaxed length: permeation is restricted to the I band. ×52,000. (From Barbosa and Pinto da Silva, 1986.)

The concept of label-fracture is simple: by labeling the outer surface of a cell and then freeze-fracturing, one should obtain images of exoplasmic faces showing underneath the electron-dense label. This approach was part of my initial plan in 1968 to prove the splitting of biological membranes during freeze-fracture; it would have provided a means of checking that an exoplasmic fracture face corresponded to a membrane whose outer surface had been covalently labeled by ferritin. The iron core of ferritin would be difficult to see across platinum-shadowed replicas, but it would be seen over regions of the E face that, because of their concavity, were sheltered from platinum shadowing. As it turned out, I did not use this approach because it was easier to see ferritin as shadowed globular molecules on etched surfaces of the membrane. A couple of years later, when labeling of cell surfaces started in earnest, the approach was again abandoned because, to obtain images of the surface distribution of the label, it would be necessary to work with uninformative, pure carbon replicas. Freeze-etching cytochemistry and platinum replicas of labeled, dried cells were better alternatives (Smith and Revel, 1972; Rosenblith et al., 1973).

Colloidal gold changed all that because of its electron opacity. While attempting to see whether colloidal gold particles could be dragged across the outer half of the membrane and be detected on protoplasmic faces, we observed that an extraordinary density of gold spheres was consistently associated to the exoplasmic faces. The replica background was littered with colloidal gold, but the images of the exoplasmic faces immediately recalled the old project. We repeated the experiment, this time without treating the replicas with sodium hypochlorite. Over a new clean background, E faces were specifically labeled; as expected, P faces were obscured by the undigested cells attached to them.

B. Technique

The sequence of steps in label-fracture is shown in Figure 13. Cells are labeled (fixed or not), impregnated (or not) in a cryoprotectant, frozen in Freon 22 (or rapid-frozen), and freeze-fractured. After thawing, the replicas are not treated with bleach or with acids. They are only washed with distilled water (if cells are not fixed before freezing, controls using fixative as thawing solutions must be performed). The typical label-fracture image provides a superimposition, in a single, coincident image, of the high-resolution conventional image of the exoplasmic fracture face and the high-resolution surface distribution of the label attached to receptors or antigens at the cell surface (Figures 14–18).

Label-fracture is a surface-labeling technique. It can be used with isolated cells or cells in monolayers (use a sandwich device for complementary replica; do not digest either side) (Boonstra et al., 1985). It can also be used with isolated epithelia (see below). In principle, it can be used to label isolated organelles (Van Berkel et al., 1986; see Figure 18), but in this case it may have to contend with the problem of adsorption of contaminants that may plague biochemical fractionation techniques.

C. The Images

The gold label is underneath the exoplasmic face (see Figure 16 for a stereoview), attached to the undigested outer half of the membrane; therefore, contrary to fracture-label, the colloidal gold spheres are not shadowed. At the edges of the fracture face, however, gold particles can be shadowed; they are at the interface of fracture. In erythrocyte membranes these shadowed gold particles are always just outside the fracture face at a distance that corresponds to the thickness of the outer half of the membrane, including the short projections of the glycoproteins being labeled (Figure 14).

D. Direct Correlation of the Surface Label to the Freeze-Fracture Morphology

For the first label-fracture experiments we located Con A and WGA receptors on human erythrocyte membranes (Pinto da Silva and Kan, 1984). Since in erythrocyte membranes intramembrane particles are not regionalized, it was not possible to relate the label to the particles (Figure 14). In rat thymocytes incubated before fixation in phosphate-buffered saline (37°C, 30 minutes) to aggregate the intramembrane particles, we established a direct correlation between particles and WGA receptors. The two domains (intramembrane particles and colloidal gold spheres) were almost coincident, a finding that attested to the high resolution of label fracture.

We studied next the microdistribution of Con A and WGA surface receptors in boar sperm cells (Figure 15). Label-fracture is an ideal technique to illustrate—close to the molecular level—the distribution of these sites and to relate them to the complex freeze-fracture morphology of the sperm plasma membrane. I refer here only to some findings concerning the base of the boar sperm head and adjacent neck areas (Kan and Pinto da Silva, in press). WGA receptors are uniformly distributed over the acrosomal area and then decrease gradually toward the base. This gradual decrease in density appears to parallel that of the large intramembrane particles, but dispersion of the latter prevents direct correlation. Along a rim at the base of the head, the density of WGA receptors increases abruptly; this area is strikingly free of particles. This particle-free area is not labeled by Con A. Over the neck, the intramembrane particles generally appear aggregated; here the particle-rich areas (but not the particle-free regions) are well labeled by Con A, whereas WGA labels both particle-rich and particle-free regions.

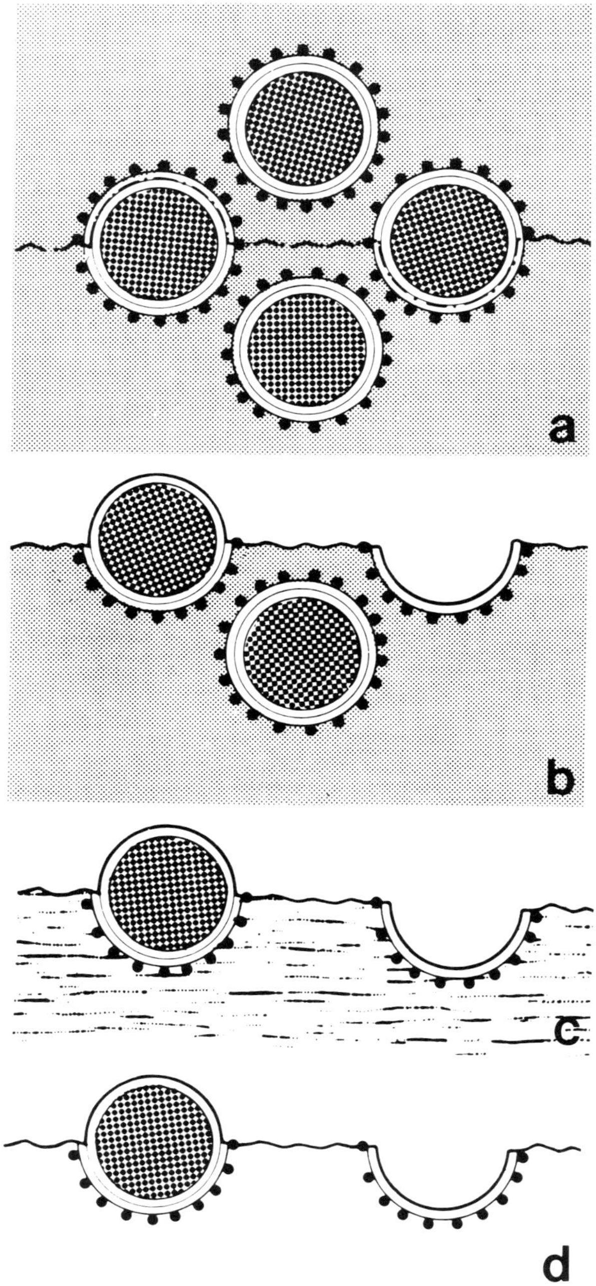

Figure 13

We have started cytochemical characterization of the apical surface of toad bladder epithelial cells. Our aim is to label the characteristic aggregates of membrane particles that are induced by vasopressin and thought to represent the structures responsible for water transport (Chevalier et al., 1986). Incubation of unfixed epithelia in WGA induces aggregation of the numerous large particles that, in these cells, partition with the exoplasmic half of apical regions of the plasma membrane. Observation of stereo pairs of label-fractured cells shows that the WGA receptors are associated with an extracellular matrix that corresponds to the very long extracellular projections of the components in the large intramembrane particles (Figure 16), as noted above. The length of these projections leads to parallax errors in nonstereo images, e.g., leads to an apparent mismatching of the distribution of label and that of the particles. The typical particle aggregates induced by vasopression or by cAMP (adenosine 3′:5′-cyclic monophosphate) do not appear to be labeled by WGA (Chevalier et al., 1986).

E. Prospects

The high resolution of label-fracture pushes it close to the molecular level. There are still problems to solve. These relate to ruling out alterations in the microdistribution of the label during thawing, washing, and drying of the replica, although there are not at present any instances where we suspect that this may have happened. As established by fracture-label, transmembrane protein molecules may partition *in toto* across the outer half of the membrane. Therefore, label-fracture may not label all transmembrane proteins since it appears that colloidal gold spheres are not pulled through the exoplasmic half of the membrane. In these instances it is possible

---

Figure 13. Label-fracture. (a) Cells in suspension are labeled and frozen. (b) Freeze-fracture splits the plasma membranes of labeled cells into exoplasmic halves (with attached surface label; *right*) and protoplasmic halves (which remain attached to the cell body; *left*). Pt/C evaporation produces a high-resolution cast of the fractured cells (only label at the interface of fracture is exposed and shadowed.) (c) Fractured, shadowed specimens are thawed and washed with distilled water, removing unfractured cells. Exoplasmic membrane halves remain attached to the replica. Coincident images of the Pt/C replica of the E face and the surface label are produced. Cells with fractured P faces (*left*) remain attached to the replica (but electron density of the cell body prevents observation of the P face). (d) The replicas are then mounted on electron microscope grids, dried, and observed. (From Pinto da Silva and Kan, 1984.)

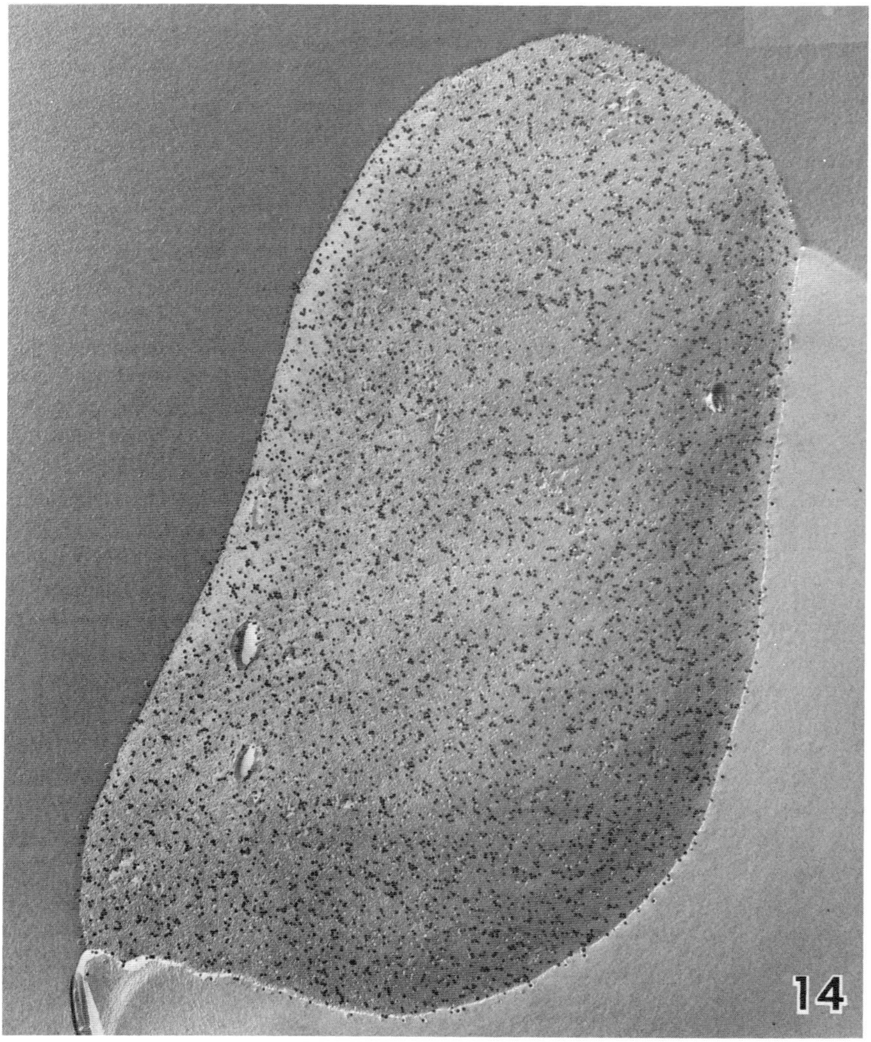

Figure 14. WGA label-fracture of human erythrocyte membranes. The label is confined to within 15 nm of the edge of the exoplasmic face. ×38,000. (From Pinto da Silva and Kan, 1984.)

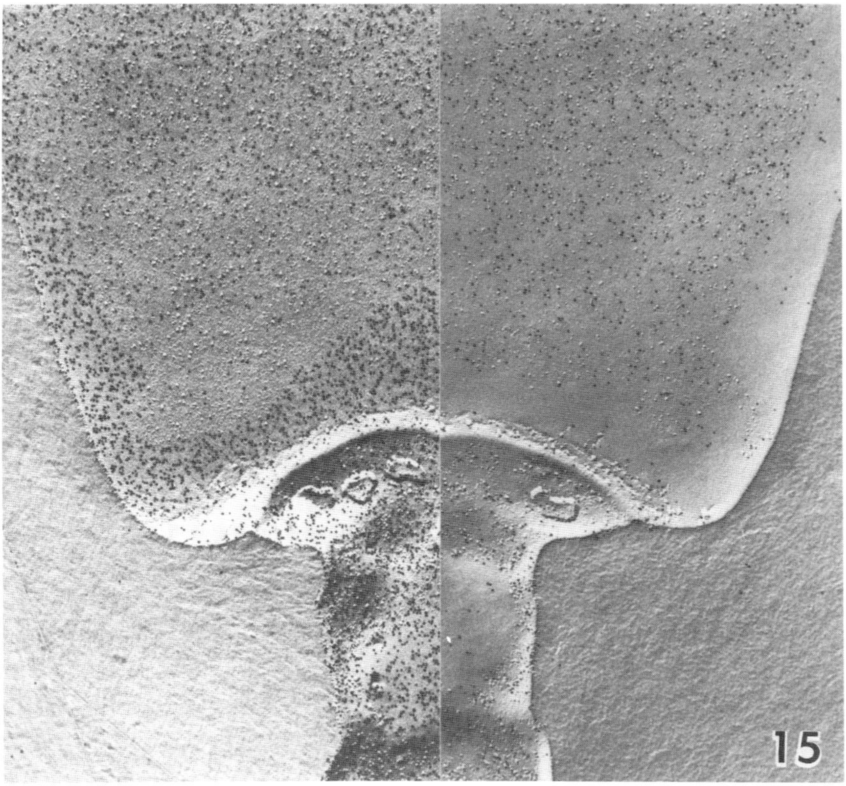

Figure 15. Label-fracture of boar spermatozoa. *Left*: WGA label-fracture of surface of boar sperm. WGA labeling is seen over the acrosomal region decreasing toward the neck. Over the particle-free zone at the base of the head, an unusually high concentration of WGA receptors is observed. The striated cords below this zone are not labeled. Over the tail, WGA labels uniformly both particle-free areas and intramembrane particle-rich areas. *Right*: Con A label-fracture of boar sperm. High density of Con A labeling is observed over acrosomal as well as the postacrosomal region of the sperm head. In contrast to WGA, Con A fails to label the particle-free zone. In the midpiece of the tail, Con A labeling is strong and appears to be related to the distribution of intramembrane particles. Both, ×44,000. (Adapted from Kan and Pinto da Silva, 1987, in press.)

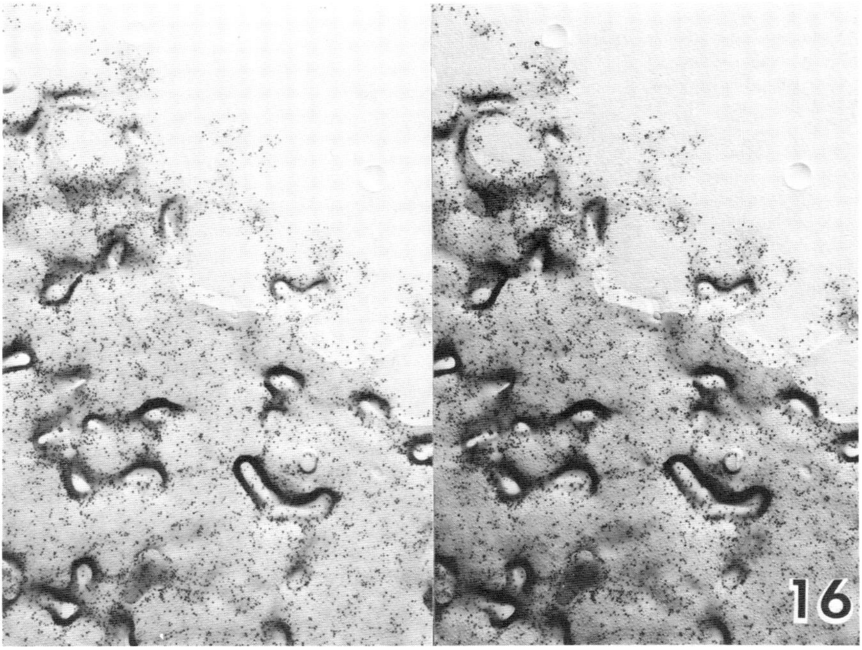

Figure 16. Stereoscopic pair showing label-fractured frog urinary bladder. Gold particles appear under the fracture face at a distance that corresponds to the extension of glycoprotein molecules into the external medium of the cell. (From Chevalier et al., 1986.)

that the gold particle will separate from transmembrane proteins and diffuse away during thawing and washing. I believe that label-fracture has the potential to become the standard technique for high-resolution labeling of cell surfaces. It is a paradox that while labeling of freeze-fractured membranes is achieved, by fracture-label, without resort to conventional (and expensive, and complex) freeze-fracture equipment, high-resolution labeling of cell surfaces will require expertise in conventional freeze-fracture and the interpretation of its images.

The reader can now imagine the really large number of problems that can be approached (better, reapproached!). He/she can also foresee the number of variants that are open for anyone to explore. The recent adaptation to autoradiography (Figure 17) and the use of cell monolayers and isolated organelles (Figure 18) are but the first examples (Carpentier et al., 1985; Boonstra et al., 1985; Van Berkel et al., 1986).

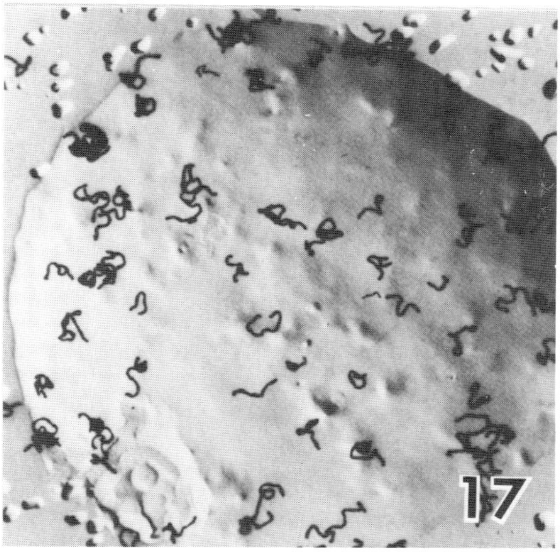

Figure 17. Lable-fracture autoradiography of a HL-60 myelomonocytic cell incubated for 30 minutes at 4°C in the presence of ^{125}I-transferrin. x9,000. (From Carpentier et al., 1985.)

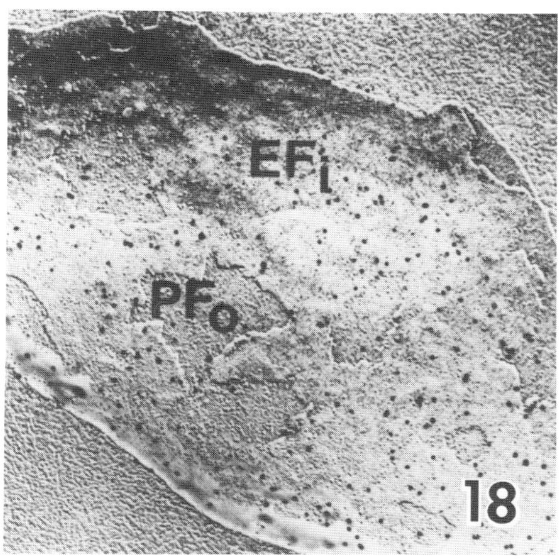

Figure 18. Immunogold label-fracture of the $M_r^{22,000}$ polypeptide in isolated, glutaraldehyde-fixed spinach chloroplasts. ×33,000. (From Van Berkel et al., 1986.)

VI. CONCLUDING REMARKS

I believe that freeze-fracture is revealing itself as the least damaging method to expose receptors and antigens for labeling. Its random course ensures the steric availability of plasma and intracellular membranes, of cytoplasm and nucleoplasm components, as well as of extracellular matrices. In addition, fracture-label with monoclonal antibodies to adequate epitopes should locate sites structurally buried within the membrane interior. The exposure of all of these components is genuine—i.e., they are not embedded within polymerized matrices and, therefore, are sterically available to the label. The continued expansion of cytochemical techniques based on freeze-fracture, as well as of those based on cryosectioning or direct labeling of plastic-embedded sections, will lead to other labeling methods. These will continue to bring together biochemical, immunochemical, and ultrastructural approaches to study the biology of the cell.

ACKNOWLEDGMENTS

I am grateful to Drs. Arthur Aguas and Frederick Kan for numerous discussions and suggestions, and to the Central Clerical Pool for their editorial skill in preparation of the manuscript. We thank the following publishers for permission to reproduce micrographs originally printed in their publications: The Rockefeller University Press; Elsevier Science Publishers B.V.; The National Academy of Sciences; Alan R. Liss Inc.; The Histochemical Society Inc.; Springer-Verlag.

NOTES

1. The image of a thin section is the integral of the electron density across the specimen. As the reconstituted bilayered structures are random, depending on the "mesh" of the reorganization, one sees trilaminar profiles with varying degrees of electron density. Should the mesh be much smaller than the section thickness, a continuous (but attenuated) trilaminar profile should be seen. On the other hand, in membranes where transmembrane proteins occupy much of the plane of the membrane, the formation, postfracture, of an interrupted bilayered structure may not occur. This appears to be the case in retinal rod disk membranes where the thin-section profiles of fractured membranes do not appear different from those in unfractured specimens (De Foe and Besharse, 1985).

2. In fracture-labeled *A. castellani* cells, we found one possible source of artifact: upon cross-fracture of the cytoplasm most granules diffused out and some attached to protoplasmic and exoplasmic fracture faces. As the glycogen granules can be identified both in sections and in Pt/C replicas, their contamination of fracture faces was only a nuisance. In addition, we have frequently observed that, when membranes are cross-fractured, small vesicles are seen forming at the cross-fracture. We believe that these vesicles are formed by membrane lipids that are not fixed by glutaraldehyde. We have seen also, in cross-fractured myelin, that the fractured edges of adjoining cross-fractured membranes become mutally sealed into a continuous structure (A. P. Aguas and P. Pinto da Silva, unpublished observations).

3. A technical snag that must and will be resolved is the fragmentation of fracture-label replicas during digestion and washing. This can be a time-consuming and frequently frustrating procedure. We have tried rotary carbon shadowing, extra carbon, and recently collodion reinforcement. For the moment it appears that careful handling, patience, and the recovery of even tiny fragments of the replica are the best approach. Technical difficulties have been the norm in the development of electron microscopic techniques but have generally been resolved.

REFERENCES

Aguas, A. P., and Pinto da Silva, P. (1983) Regionalization of transmembrane glycoproteins in the plasma membrane of boar sperm head as revealed by fracture-label. J. Cell Biol. *97:* 1356–1364.

Aguas, A. P., and Pinto da Silva, P. (1984) High density of transmembrane glycoproteins on the flagellar surface of boar sperm cells. J. Cell Biol. *99:* 655–660.

Aguas, A. P., and Pinto da Silva, P. (1985) The acrosomal membrane of boar sperm: A Golgi derived membrane poor in glycoconjugates. J. Cell Biol. *100:* 528–535.

Bakayev, V. V., Schmatchenko, V. V., and Georgiev, G. P. (1979) Subnucleosome particles containing high mobility group proteins HMG-E and HMG-G originate from transcriptionally active chromatin. Nucleic Acids Res. *7:* 1525–1540.

Barbosa, M. L. F., and Pinto da Silva, P. (1983) Restriction of glycolipids to the outer half of a plasma membrane: Concanavalin A labeling of membrane halves in *Acanthamoeba castellani.* Cell *33:* 959–966.

Barbosa, M. L. F., and Pinto da Silva, P. (1985) Fracture-permeation: Assessment of intermolecular spaces in the cytoplasm. J. Cell Biol. *101:* 408a.

Barbosa, M. L. F., and Pinto da Silva, P. (1986a) Ultrastructural patterns of ferritin permeation into glutaraldehyde-fixed sarcomeres characterize stages of contraction in striated muscle. J. Electron Microsc. Tech. *4:* 329–342.

Barbosa, M. L. F., and Pinto da Silva, P. (1986b) Fracture-permeation: A technique to assess cytoplasmic compactness after glutaraldehyde fixation. J. Electron Microsc. Tech. *4:* 385–397.

Benson, A. A., and Singer, S. J. (1965) The lipoprotein model for membrane structure. Abstracts, 150th National Meeting, Am. Chem. Soc., Atlantic City, September, p. 8c.

Bloodgood, R. A. (1981) The flagellum as a model system for studying dynamic cell surface events. Cold Spring Harbor Symp. Quant. Biol. *46:* 683–693.

Boonstra, J., Van Belzen, N., Van Maurik, P., Hage, W. J., Blok, F. J., Wiegant, F. A. C., and Verkleij, A. J. (1985) Immunocytochemical demonstrations of cytoplasmic and cell surface EGF receptors in A431 cells using cryo-ultramicrotomy, surface replication, freeze-etching and label fracture. J. Microsc. *140:* 119–129.

Bowers, B. (1980) A morphological study of plasma and phagosome membranes during exocytosis in *Acanthamoeba.* J. Cell Biol. *84:* 246–260.

Branton, D. (1966) Fracture faces of frozen membranes. Proc. Natl. Acad. Sci. US *55:* 1048–1056.

Branton, D., and Moor, H. (1964) Fine structure in freeze-etched *Allium cepa* L. root tips. J. Ultrastruct. Res. *11:* 401–411.

Carpentier, J.-L., Brown, D., Iacopetta, B., and Orci, L. (1985) Detection of surface bound ligands by freeze-fracture autoradiography. J. Cell Biol. *101:* 887–890.

Chevalier, J., Pinto da Silva, P., Ripoche, P., Gobin, R., Wang, X. Y., Grossetete, J., and Bourguet, J. (1985) Structural and cytochemical differentiation of membrane elements of the apical membrane of amphibian bladder epithelial cells. A label-fracture study. Biol. Cell *55:* 181–190.

Danielli, J. F. (1954) The present position in the field of facilitated diffusion and selective active transport. Proc. Symp. Colston Res. Soc. *7:* 1–14.

De Foe, D. M., and Besharse, J. C. (1985) Membrane assembly in retinal photoreceptors. II. Immunocytochemical analysis of freeze-fractured rod photoreceptor membranes using anti-opsin antibodies. J. Neurosci. *5:* 1023–1034.

Fisher, K. A., and Yanagimoto, K. C. (1986) Effect of membrane splitting on transmembrane polypeptides. J. Cell Biol. *102:* 551–559.

Friend, D. S. (1982) Plasma membrane diversity in a highly polarized cell. J. Cell Biol. *93:* 243–249.

Green, D., and Perdue, J. F. (1966) Membranes as expressions of repeating units. Proc. Natl. Acad. Sci. USA *55:* 1295–1302.

Guatelli, J. C., Porter, K. R., Anderson, K. L., and Boggs, D. P. (1982) Ultrastructure of the cytoplasmic and nuclear matrices of human lymphocytes observed using high voltage electron microscopy of embedment-free sections. Biol. Cell *43:* 69–80.

Kan, F. W. K., and Pinto da Silva, P. (1986) Preferential association of glycoproteins to the enchromatin regions of cross-fractured nuclei is revealed by fracture-label. J. Cell Biol. *102:* 576–586.

Kan, F. W. K., and Pinto da Silva, P. (in press) Molecular demarcation of surface domains as established by label-fracture cytochemistry of boar spermatozoa. J. Histochem. Cytochem.

Korn, E. D., and Wright, P. L. (1973) Macromolecular composition of an amoeba plasma membrane. J. Biol. Chem. *248:* 439–447.

Korn, E. D., Dearborn, D. G., and Wright, P. L. (1974) Lipophosphonoglycan of the plasma membrane of *Acanthamoeba castellanii*: Isolation from whole amoebas and identification of the water-soluble products of acid hydrolysis. J. Biol. Chem. *249:* 3335–3341.

Levy-Wilson, B., Connor, W., and Dixon, G. H. (1979) A subset of trout testis nucleosomes enriched in transcribed DNA sequences contains high mobility group proteins as major structural components. J. Biol. Chem. *254:* 609–620.

Marchesi, V. T., Tillack, T. W., Jackson, R. L., Segrest, J. P., and Scott, R. D. (1972) Chemical characterization and surface orientation of the major glycoprotein of the human erythrocyte membrane. Proc. Natl. Acad. Sci. USA *69:* 1445–1449.

Moor, H., and Muhlethäler, K. (1963) Fine structure of frozen-etched yeast cells. J. Cell Biol. *17:* 609–628.

Pinto da Silva, P. (1973) Membrane intercalated particles in human erythrocyte ghosts: Sites of preferred passage of water molecules at low temperature. Proc. Natl. Acad. Sci. USA *70:* 1339–1343.

Pinto da Silva, P. (1986a). Freeze-fracture: Seeing and thinking biological membranes. In *Biomembranes and Receptor Mechanisms* (E. Bertoli, ed.). Fidia Research Series, Padua, Italy.

Pinto da Silva, P. (1986b). Freeze-fracture cytochemistry: A poor person's guide. Bull. Electron Microsc. Soc. Am. (forthcoming).

Pinto da Silva, P., and Branton, D. (1970) Membrane splitting in freeze-etching: Covalently bound Ferritin as a membrane marker. J. Cell Biol. *45:* 598–605.

Pinto da Silva, P., and Kan, F. W. K. (1984) Label-fracture: A method for high resolution labeling of cell surfaces. J. Cell Biol. *99:* 1156–1161.

Pinto da Silva, P., and Nicolson, G. L. (1974) Freeze-etch localization of concanavalin A receptors to the membrane intercalated particles in human erythrocyte membranes. Biochim. Biophys. Acta *363:* 311–319.

Pinto da Silva, P., and Torrisi, M. R. (1982) Freeze-fracture cytochemistry: Partition of glycophorin in freeze-fractured erythrocyte membranes. J. Cell Biol. *93:* 463–469.

Pinto da Silva, P., Douglas, S. D., and Branton, D. (1970) Localization of A-1 antigens on the human erythrocyte membrane. J. Cell Biol. *47:* 159a.

Pinto da Silva, P., Douglas, S. D., and Branton, D. (1971) Localization of A antigen sites on human erythrocyte ghosts. Nature (Lond.) *232:* 194–196.

Pinto da Silva, P., Moss, P. S., and Fudenberg, H. H. (1973) Anionic sites on the membrane: Intercalated particles of human erythrocyte ghost membranes; freeze-etch localization. Exp. Cell Res. *81:* 127–138.

Pinto da Silva, P., Kachar, B., Torrisi, M. R., Brown, C., and Parkison, C. (1981a) Freeze-fracture cytochemistry: Replicas of critical point dried cells and tissues after "fracture-label." Science (Wash. DC) *213:* 230–233.

Pinto da Silva, P., Parkison, C., and Dwyer, N. (1981b) Fracture-label: Cytochemistry of freeze-fracture faces in the erythrocyte membrane. Proc. Natl. Acad. Sci. USA *78:* 343–347.

Pinto da Silva, P., Parkison, C., and Dwyer, N. (1981c) Freeze-fracture cytochemistry. II. Thin sections of cells and tissues after labelling of fracture faces. J. Histochem. Cytochem. *29:* 917–928.

Pinto da Silva, P., Torrisi, M. R., and Kachar, B. (1981d) Freeze-fracture cytochemistry: Localization of wheat germ agglutinin and concanavalin A binding sites on freeze-fractured pancreatic cells. J. Cell Biol. *91:* 361–372.

Pinto da Silva, P., Barbosa, M. L. F., and Aguas, A. P. (1986) A guide to fracture-label: Cytochemical labeling of freeze-fractured cells. In *Advanced Techniques in Biological Microscopy* (I. K. Koehler, ed.), Vol. 3, pp. 201–227. Springer-Verlag, New York.

Porter, K. R., Boggs, D. P., and Anderson, K. L. (1982) The distribution of water in the cytoplasm. Annals of the 40th Annual Electron Microscopy Soc. Am. meeting, Washington, DC, pp. 4–7.

Primakoff, P., and Myles, D. G. (1983) A map of the guinea pig sperm surface constructed in the monoclonal antibodies. Dev. Biol. *98:* 417–428.

Pryzwansky, K. B., Schliwa, M., and Porter, K. R. (1983) Comparison of the three-dimensional organization of unextracted and Triton-extracted human neutrophilic polymorphonuclear leukocytes. Eur. J. Cell Biol. *30:* 112–125.

Reeves, R., Cheng, D., and Chung, S. C. (1981) Carbohydrate modifications of the high mobility group proteins. Proc. Natl. Acad. Sci. USA *78:* 6704–6708.

Robertson, J. D. (1961) The unit membrane. In *Electron Microscopy in Anatomy* (J. D. Boyd, F. R. Johnson, J. D. Lever, eds.), pp. 74–79. Williams & Wilkins, Baltimore.

Rosenblith, J. Z., Ukena, T. E., Yin, H. H., Berlin, R. D., and Karnovsky, M. J. (1973) A comparative evaluation of the distribution of Con A binding sites on the surfaces of normal, virally transformed and protease treated fibroblasts. Proc. Natl. Acad. Sci. USA *70:* 1625–1629.

Seve, A. P., Douvier, H. D., Masson, C., Gerand, G., and Bouteille, M. (1984) In size distribution in different cell types of nuclear glycoconjugates detected by two lectins. J. Submicrosc. Cytol. *16:* 631–641.

Sjöstrand, F. (1963) A comparison of plasma membranes, cytomembranes and mitochondrial membrane elements with respect to ultrastructural features. J. Ultrastruct. Res. *9:* 561–580.

Smith, S. B., and Revel, J.-P. (1972) Mapping of concanavalin A binding sites on the surface of several cell types. Dev. Biol. *27:* 434–441.

Tillack, T. W., Scott, R. E., and Marchesi, V. T. (1972) The structure of erythrocyte membranes studied by freeze-etching. II. Focalization of receptors for phytohemagglutinin and influenza virus to the intramembranous particles. J. Exp. Med. *135:* 1209–1221.

Torrisi, M. R., and Bonatti, S. (1985) Immunocytochemical study of the partition and distribution of Sindbis virus glycoproteins in freeze-fractured membranes of infected baby hamster kidney cells. J. Cell Biol. *101:* 1300–1306.

Torrisi, M. R., and Pinto da Silva, P. (1982) T lymphocyte heterogeneity: WGA labeling of transmembrane glycoproteins. Proc. Natl. Acad. Sci. USA *79:* 5671–5674.

Torrisi, M. R., and Pinto da Silva, P. (1984) Compartmentalization of intracellular membrane glycocomponents is revealed by fracture-label. J. Cell Biol. *98:* 29–34.

Van Berkel, J., Steup, M., Völker, W., Robenek, H., and Flügge, V. I. (1986) Polypeptides of the chloroplast membranes as vizualized by immunochemical techniques. J. Histochem. Cytochem. *34:* 577–583.

Weisbrod, S., Groudine, M., and Weintraub, H. (1980) Interaction of HMB 14 and 17 with actively transcribed genes. Cell *19:* 289–301.

Wolosewick, J. J., and Porter, K. R. (1979) Microtrabecular lattice of the cytoplasm ground substance: Artifact or reality? J. Cell Biol. *82:* 114–139.